LIGHT&SKIN
INTERACTIONS

LIGHT&SKIN
INTERACTIONS

Simulations for Computer Graphics Applications

Gladimir V. G. Baranoski

Aravind Krishnaswamy

ELSEVIER

AMSTERDAM • BOSTON • HEIDELBERG • LONDON
NEW YORK • OXFORD • PARIS • SAN DIEGO
SAN FRANCISCO • SINGAPORE • SYDNEY • TOKYO
Morgan Kaufmann Publishers is an imprint of Elsevier

MORGAN KAUFMANN

Morgan Kaufmann Publishers is an imprint of Elsevier
30 Corporate Drive, Suite 400, Burlington, MA 01803, USA

This book is printed on acid-free paper.

Notices
Knowledge and best practice in this field are constantly changing. As new research and experience
broaden our understanding, changes in research methods, professional practices, or medical treatment
may become necessary.

Practitioners and researchers must always rely on their own experience and knowledge in evaluating
and using any information, methods, compounds, or experiments described herein. In using such
information or methods they should be mindful of their own safety and the safety of others, including
parties for whom they have a professional responsibility.

To the fullest extent of the law, neither the Publisher nor the authors, contributors, or editors, assume
any liability for any injury and/or damage to persons or property as a matter of products liability,
negligence or otherwise, or from any use or operation of any methods, products, instructions, or ideas
contained in the material herein.

Library of Congress Cataloging-in-Publication Data
Application submitted.

British Library Cataloguing-in-Publication Data
A catalogue record for this book is available from the British Library.

ISBN: 978-0-12-375093-8

For information on all Morgan Kaufmann publications
visit our Web site at *www.mkp.com* or *www.elsevierdirect.com*

Typeset by: diacriTech, India

Printed in China

10 11 12 13   5 4 3 2 1

Working together to grow
libraries in developing countries

www.elsevier.com | www.bookaid.org | www.sabre.org

ELSEVIER   BOOK AID
             International   Sabre Foundation

Acknowledgements

First and foremost, we would like to thank our families for their unconditional support during the journey that led to this book.

We are also grateful to Manuel Menezes de Oliveira Neto, Jon Rokne, Steve Cunningham, Francisco Imai, Rui Bastos and Min Chen for their constructive suggestions on the first draft version, and to Paulo Alencar and Tenn Francis Chen for their valuable feedback during the late stages of this project.

One of the key tasks involved in the preparation of this book involved the gathering of data and images to illustrate important issues related to the simulation of light and skin interactions. Hence, we also would like to thank researchers and organizations that directly and indirectly contribute to this effort.

The production of this book would not be possible without the dedication and hard work of the Elsevier team involved in this project, notably Gregory Chalson, Heather Scherer and Andre Cuello. We are certainly indebted to them.

Last, but not least, we would like to acknowledge the encouragement of our colleagues from the Natural Phenomena Simulation Group (University of Waterloo) and the Visual Computing Lab (Adobe Systems Incorporated).

Contents

List of Figures

Nomenclature

Acronyms

Introduction

1

Skin is one of the most complex organs in the human body. Its appearance, determined by genetic (endogenous) and environmental (exogenous) factors, can convey important information about an individual's origin, age, and health. For this reason, it has been an object of extensive study in life and physical sciences. Several clinical procedures used to assist the diagnosis and treatment of dermatological diseases are based on the measurement and modeling of skin appearance attributes. Computational (in silico) simulations of biophysical processes affecting skin appearance are also being routinely employed, for example, in the cosmetics and entertainment industries. In the former, the predictions of such simulations are used to assist the development of skin-related products such as cosmetics and protective lotions. In the latter, they are used to generate realistic images of human features depicted in movies and games.

Among the different environmental stimuli that can affect the appearance of human skin, light triggers the most commonly observed responses. This book examines the simulation of these responses from a computer graphics point of view, i.e., an emphasis is given to the modeling of skin appearance for realistic image synthesis purposes. Skin images generated using computer graphics techniques can be loosely classified as believable or predictable. Believable images result from biophysically inspired simulations. They are usually esthetically pleasing and realistic. Predictable images, on the other hand, result from biophysically based simulations whose accuracy is evaluated through comparisons with experimental data. Predictability, however, does not guarantee the generation of images with a higher degree of realism. This may be affected by errors introduced in other components of rendering systems designed for the generation of synthetic pictures, henceforth referred to as image synthesis pipelines.

Although correct biophysical simulations are not necessarily required for the generation of convincing images, the fact that these simulations can be used in a predictive manner has important practical implications. For example, predictive models of light and skin interactions can be employed in medical investigations to derive biochemical and biophysical properties of skin specimens from in-situ noninvasive light reflection and transmission measurements. For this reason, one of the goals of this book is to provide a foundation for the development of comprehensive and predictive simulation frameworks, which can be used to model not only the interactions of light with human skin but also with other organic and inorganic materials.

For decades, scientists from different fields have been studying the photobiological processes associated to skin appearance. As a result, a large amount of relevant knowledge and data are scattered through various scientific domains. This information is organized in a concise and consistent manner in this book. The objective is to make it readily accessible by researchers, students, and practitioners interested in the simulation of these phenomena regardless of their original field of expertise. Ultimately, this book aims to foster the cross-fertilization and synergistic collaboration among different scientific communities working on the same problem, namely the simulation of light and skin interactions.

The following paragraphs describe the organization of the book, which is divided into three main modules. In the first module, Chapters 2–5, we provide a biophysical substrate for the development and evaluation of predictive models of light interaction with human skin. In the second module, Chapters 6–8, we introduce the different modeling approaches used by computer graphics researchers to render the appearance of human skin. We group these approaches, and their representative models, according to the main simulation guideline that completely distinguishes each family of models from the other. In the final module, Chapters 9 and 10, we address specific practical issues and open problems related to the simulation of light and skin interactions.

Chapter 2 presents key physical concepts and terminology employed in the modeling of skin appearance. We start by examining the nature of light and providing a review of relevant optics laws. This presentation is followed by a concise description of fundamental light and matter interaction processes: emission, scattering, and absorption. We also define tissue optics terms and physical quantities normally used to describe the appearance of different materials.

Chapter 3 gives a description of the image synthesis context where skin appearance models are inserted. We outline fundamental approaches used to simulate global and local light transport. We also describe simulations of virtual measurement devices that can be employed to evaluate the accuracy of local lighting models before they are incorporated into an image synthesis pipeline. The final stage in this pipeline corresponds to the mapping of simulated physical quantities to display (or printing) values to generate an image. An overview of the main steps of this mapping closes this chapter.

To simulate the light interaction processes associated with the appearance of human skin correctly, it is necessary to understand the inherent complexities of this multilayered and inhomogeneous organ. For this purpose, in Chapter 4, we examine the biophysical and structural characteristics of the various skin tissues and how these characteristics affect the processes of light propagation and absorption that determine the color, glossiness, and translucency of human skin. Several photobiological processes, such as tanning, are triggered by electromagnetic radiation outside the visible portion of the light spectrum. We also look at these light interactions and examine their role in scientific applications such as the visual diagnosis of skin diseases.

A substantial amount of research on the simulation of skin and light interactions is performed in fields as diverse as biomedical optics, colorimetry, and remote sensing. In Chapter 5, we provide a review of relevant models developed in these fields and organize them according to the simulation approach used in their development. Any scientifically sound effort to simulate skin appearance has to carefully account for this valuable body of interdisciplinary work. In fact, several models aimed at the synthesis of realistic images of human skin were built on simulation frameworks developed outside the computer graphics domain.

Chapter 6 presents biologically inspired models employed in the realistic rendering of skin appearance. The algorithms used by these models were designed to reproduce some of the visual characteristics of human skin and contribute to the generation of believable images. Although predictability was not one of their design guidelines, these models introduced to the graphics community several skin biology concepts and set the stage for the development of a new generation of predictive models.

Chapter 7 examines a comprehensive model developed using a first-principles approach. More specifically, this model simulates the processes of light propagation and absorption within human skin, taking into account the specific properties of relevant tissue constituents. It was one the first models in this area to have the fidelity of its simulations evaluated through qualitative and quantitative comparisons of predicted results with actual measured data. For this reason, we perform a detailed review of the procedures and data used in its evaluation.

Chapter 8 discusses a family of models that employs light transport algorithms based on the application of the diffusion theory approach. This theory, outlined in Chapter 4, provides an approximated solution for the general equation used to describe light propagation in optically turbid media. The first models developed using this approach were aimed at the generation of believable images, i.e., the evaluation of their predictions was based on the visual inspection of the resulting images. The subsequent incremental refinement of their algorithms led to the design of more comprehensive models, which was followed by the use of more reliable evaluation procedures.

In Chapter 9, we examine key challenges related to the development and deployment of predictive models of light and skin interactions. We address practical issues, such as data scarcity, and outline alternatives to reduce the computational costs associated with the incorporation of these models into efficient realistic image synthesis pipelines.

Finally, in Chapter 10, we discuss open problems related to simulation of light and skin interactions and the realistic rendering of skin appearance. We also look at future prospects involving the use of these simulations to support theoretical and applied research beyond the computer graphics domain.

This is the first book devoted to the simulation of light interactions with human skin. Its focal point is the modeling of physical quantities that describe skin appearance attributes. Other topics related to the rendering of human features, such as the geometric modeling and texture mapping of skin surfaces, are beyond its scope.

As computer graphics researchers, we enjoy all aspects of the image synthesis process, and we fully appreciate the stimulating value of beautiful images. We are also fully aware of the time and effort required to obtain images that not only look right but are also biophysically correct. We have been involved in several projects leading to this goal, and this book is meant to assist students, researchers, and practitioners on a similar undertaking aimed at the rendering of material appearance. Furthermore, we believe that such a journey does not necessarily end after the generation of realistic images. As mentioned earlier, a sound and well-designed model can be a valuable tool in different scientific domains. We address these ideas in this book, and we hope that their discussion will contribute to further extend the computer graphics horizons.

Light, optics, and appearance

2

In this chapter, we provide the physical background for the discussion of the processes of light interaction with human skin. We start with a brief discussion of the nature of light and a concise review of fundamental optics concepts. This presentation is then followed by an overview of the main light and matter interaction processes and the physical terms usually employed in their quantification.

The interactions of light with human tissue are the object of extensive investigations in a myriad of fields, from biomedical optics to remote sensing. To maximize the cross-fertilization with these fields, it is necessary to use a consistent terminology. For this purpose, in this chapter, we also define relevant tissue optics terms.

Ultimately, light interactions determine the appearance attributes (e.g., color and glossiness) of a given skin specimen. Accordingly, this chapter closes with a review of the physical quantities associated with these attributes.

2.1 LIGHT AS RADIATION

Light is a form of energy, which includes not only the visible light but also other forms of electromagnetic radiation such as microwaves and x-rays. The parameter used to distinguish among the different types of radiation is the wavelength, λ, which is usually measured in nanometers (nm, 10^{-9} m) or Angstroms (Å, 10^{-10} m). More precisely, this parameter provides the spectral distribution, or spectrum, of electromagnetic radiation [169]. An abridged chart describing the electromagnetic spectrum, or spectral distribution, of light is given in Figure 2.1. Note that the spectral range of visible radiation does not have precise limits since these limits vary from person to person [169].

DOI: 10.1016/B978-0-12-375093-8.00002-2

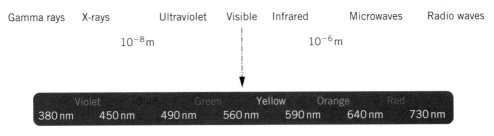

FIGURE 2.1

Abridged chart describing the electromagnetic spectrum, or spectral distribution, of light.

Isaac Newton (1643–1727) observed that visible white light could be split by a glass prism into a rainbow of colors, which could not be further subdivided. He also thought that light was made of "corpuscles." He was right, but the reasoning that he used to reach this conclusion was erroneous [86]. Later, it was discovered that light behaves like waves. Such a behavior was formalized in the classical electromagnetic theory presented by Maxwell (1831–1879). In the beginning of the twentieth century, however, it was found that light does indeed sometimes behave like a particle [87].

In 1900, Planck demonstrated that matter does not emit light continuously as it was predicted by the electromagnetic theory of Maxwell. Instead, according to Planck, light is emitted as small packets of energy, or *quanta*. This idea that energy comes only in discrete quantities is considered the beginning of quantum mechanics, a revolutionary theory for submicroscopic phenomena [111], and it is also essential for the understanding of light interactions with matter. In 1905, building on Planck's idea, Einstein postulated that energy along an incident beam is quantized into "particles," later called *photons* [156], whose individual energy is given in terms of their wavelength and Planck's constant [111].

The concept of photons is fundamental for geometrical optics [111, 159], also called ray optics, which involves the study of the particle nature of light. In geometrical optics, the large-scale behavior of light, such as reflection and refraction, is described by assuming light to be composed of noninteracting rays, each of them carrying a certain amount of energy. Also, in 1905, Schuster [218] published a paper on radiation through a foggy atmosphere, and later, in 1906, Schwarzschild published a paper on the equilibrium of the Sun's atmosphere [173]. These two astrophysical works are considered the beginning of the radiative transfer theory [15, 41, 207]. This theory combines principles of geometrical optics and thermodynamics to characterize the flow of radiant energy at large scales compared with its wavelength and during large time intervals compared to its frequency [15].

There are phenomena at the level of electromagnetism, such as interference, diffraction, and polarization, that cannot be explained by geometrical optics or radiative transfer theory. Briefly, interference refers to the phenomenon that waves, under certain conditions, intensify or weaken each other [175], diffraction corresponds to the slight bending of light that occurs when light passes very close to an edge [190], and polarization occurs when the electrical portion of the light waves moves in a single direction rather than in random directions [190]. These phenomena are addressed by physical optics [111, 159], also called wave optics, which involve the study of the wave nature of light.

Despite their limitations, geometrical optics and radiative transfer theory are the two levels of physical description largely used in the simulation of visual phenomena aimed at image synthesis applications. This is a practical choice since the simulation of phenomena such as interference, diffraction, and polarization usually requires substantial computational resources to produce results of significantly higher accuracy than those attainable by geometrical optics. From a practical point of view, it is more efficient to model light as rays rather than waves. We can think of a wave as just a ray with energy. In addition, spectral-dependent quantities can be accounted for in ray optics simulations by associating a wavelength, a wave optics parameter, with each ray.

These simulation strategies are not guided only by computational costs. There are also deeper scientific issues that need to be considered. For example, it is still not possible to fully explain partial reflection of light by two surfaces (e.g., glass), i.e., simultaneously accounting for both ray optics and wave optics phenomena (e.g., interference). The best that one can do is to calculate the correct probability that light will reflect from a surface of a given thickness. Such a calculation is performed under the framework provided by the quantum electrodynamics (QED) theory [86].

Viewed in this context, the light transport simulations performed in computer graphics and related fields (e.g., tissue optics and remote sensing) can be seen as further simplifications of this general framework, albeit, in many instances, using the same probability theory tools.

Another simplifying assumption commonly employed in light transport simulations is the decoupling of the energies associated with different wavelengths. In other words, the energy associated with some region of the space, or surface, at wavelength λ_1 is independent of the energy at another wavelength λ_2.

2.2 OPTICS CONCEPTS

The reflection and transmission (refraction) of light at the smooth surfaces of pure materials (i.e., not mixed with any other substance or material) is described by the Fresnel equations [111]. Before getting to the specifics of these equations, we shall review some relevant physical parameters, definitions, and laws.

Materials such as conductors (metals), semiconductors, and dielectrics are characterized by their complex index of refraction, N, which is composed of a real and an imaginary term. The real term corresponds to the real index of refraction, η, also called refractive index, which measures how much an electromagnetic wave slows down relative to its speed in vacuum [250]. The imaginary term corresponds to the extinction index, k, also called the extinction coefficient, which represents how easily an electromagnetic wave can penetrate into the medium [250]. The resulting expression for the complex index of refraction is given by

$$N(\lambda) = \eta(\lambda) + jk(\lambda), \qquad (2.1)$$

where j is an imaginary number.

The extinction coefficient is related to the conductive properties of the material, where semiconductors are materials with a small extinction coefficient and dielectrics are essentially nonconductors whose extinction coefficient is by definition zero [225].

Most of the quantities presented in this section may be made dependent on the wavelength λ. For example, Figure 2.2 provides a graph illustrating the wavelength dependency of the refractive index of pure water [191], while Figure 2.3 provides a graph illustrating the wavelength dependency of its extinction coefficient [197]. For notational simplicity, however, we will omit such a dependency, represented by λ, in the following definitions.

When light hits a smooth surface, its reflection direction, represented by the vector $\vec{r}$ (Figure 2.4), is obtained using the law of reflection [111]. This law states that the polar angle of the reflection direction, θ_r, is equal to the polar angle of incidence, θ_i, and will be in the same plane as the incidence direction, denoted by ψ_i and represented by the vector $\vec{i}$, and the surface normal, represented by the vector $\vec{n}$. This law is given by

$$\theta_r = \theta_i, \qquad (2.2)$$

where the angle θ_i can be obtained using the following equation:

$$\theta_i = \arccos\left(\frac{\vec{n} \cdot \vec{i}}{|\vec{n}| |\vec{i}|} \right). \qquad (2.3)$$

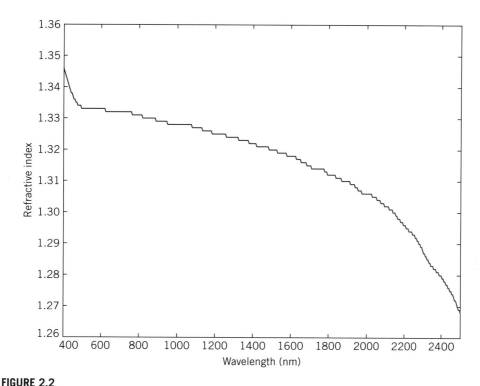

FIGURE 2.2

Refractive index of pure water (400–2500 nm) [191].

Considering the geometry described in Figure 2.4, and applying the law of reflection stated above, the reflection direction, $\vec{r}$, is given by

$$\vec{r} = \vec{i} - 2\vec{n}\cos\theta_i = \vec{i} - 2\vec{n}(\vec{i} \cdot \vec{n}). \qquad (2.4)$$

Refraction can be defined as the bending or the change in the direction of the light rays as they pass from one medium to another [119]. This bending is determined by the change in the velocity of propagation associated with the different indexes of refraction of the media [77]. The refraction (transmission) direction, represented by the vector $\vec{t}$ (Figure 2.4), is obtained using the law of refraction, also known as Snell's law [111]:

$$\eta_i \sin\theta_i = \eta_t \sin\theta_t, \qquad (2.5)$$

where η_i is the refractive index of the material (medium) of incidence and η_t is the refractive index of the transmissive material (medium).

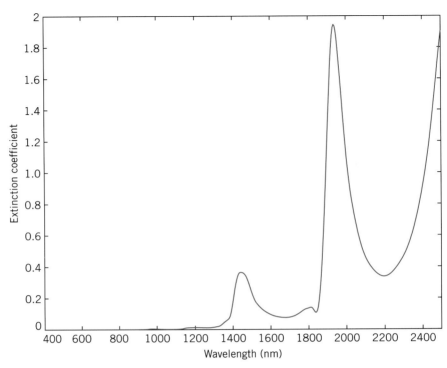

FIGURE 2.3

Extinction coefficient of pure water (400–2500 nm) [191, 197].

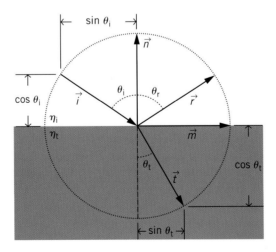

FIGURE 2.4

Geometry for light incident at an interface between different materials.

The polar angle of transmission, θ_t, becomes 90° when θ_i is equal to the "critical angle," θ_c, given by [88]

$$\theta_c = \arcsin\left(\frac{\eta_t}{\eta_i}\right). \tag{2.6}$$

At all polar angles of incidence greater than the critical angle, the incident light is reflected back to the incident medium, a phenomenon known as total internal reflection [88].

The transmission (refraction) direction $\vec{t}$ is then given by

$$\vec{t} = -\vec{n}\cos\theta_t + \vec{m}\sin\theta_t, \tag{2.7}$$

where $\vec{m}$ is the vector perpendicular to $\vec{n}$ and in the same plane as $\vec{i}$ and $\vec{n}$.

Equation 2.7 can be expanded to yield [112]

$$\vec{t} = \left[-\frac{\eta_i}{\eta_t}(\vec{i}\cdot\vec{n}) - \sqrt{1 - \left(\frac{\eta_i}{\eta_t}\right)^2\left(1 - (\vec{i}\cdot\vec{n})^2\right)} \right]\vec{n} + \frac{\eta_i}{\eta_t}\vec{i}. \tag{2.8}$$

It is worth noting that the expression inside the square root in Equation 2.8 may be negative. This indicates that total internal reflection has occurred.

The incident rays are not only reflected or transmitted (refracted) at an interface between dielectrics but also attenuated. This attenuation is given by the Fresnel coefficients for reflection and transmission (refraction), which are computed using the Fresnel equations [111]. More specifically, the Fresnel coefficients for reflection of light polarized in directions perpendicular to ($F_{R\perp}$) and parallel ($F_{R\parallel}$) to an interface can be computed using the following expressions [225]:

$$F_{R\perp} = \frac{b_1^2 + b_2^2 - 2b_1\cos\theta_i + \cos^2\theta_i}{b_1^2 + b_2^2 + 2b_1\cos\theta_i + \cos^2\theta_i} \tag{2.9}$$

and

$$F_{R\parallel} = F_{R\perp}\frac{b_1^2 + b_2^2 - 2b_1\sin\theta_i\tan\theta_i + \sin^2\theta_i\tan^2\theta_i}{b_1^2 + b_2^2 + 2b_1\sin\theta_i\tan\theta_i + \sin^2\theta_i\tan^2\theta_i}, \tag{2.10}$$

where b_1 and b_2 are given by

$$b_1^2 = \frac{1}{2\eta_i^2}\left\{\sqrt{(\eta_t^2 - k^2 - \eta_i^2\sin^2\theta_i)^2 + 4\eta_t^2 k^2} + \eta_t^2 - k^2 - \eta_i^2\sin^2\theta_i\right\} \quad (2.11)$$

and

$$b_2^2 = \frac{1}{2\eta_i^2}\left\{\sqrt{(\eta_t^2 - k^2 - \eta_i^2\sin^2\theta_i)^2 + 4\eta_t^2 k^2} - (\eta_t^2 - k^2 - \eta_i^2\sin^2\theta_i)\right\}. \quad (2.12)$$

Brewster empirically discovered that light from a surface is completely polarized if $\vec{r}$ and $\vec{t}$ form a right angle [87], which corresponds to $\theta_i + \theta_t = 90°$ [32]. In this case, we have the angle of incidence θ_i equal to the Brewster's angle θ_f [87], also called polarization angle [32], and the perpendicular term $F_{R_\perp}$ drops to zero. The Brewster's angle is given by the Brewster's law

$$\tan\theta_f = \frac{\eta_t}{\eta_i}. \quad (2.13)$$

For some materials (e.g., plant leaves), polarization is affected by surface features. Hence, the measurement of the entirely parallel-polarized reflected light at the Brewster's angle allows the decoupling of surface and subsurface light reflection behaviors. This decoupling has relevant practical and theoretical applications in life science studies such as those involved in the remote sensing of vegetation [104, 162].

The Fresnel coefficient for reflection, or reflectivity [175], F_R, for polarized light is the weighted sum of the polarized components, in which the weights must sum to one [250]. For unpolarized light, the Fresnel coefficient for reflection is simply the average of the two coefficients $F_{R_\perp}$ and $F_{R_\|}$. Then, the equation used to compute this coefficient for dielectrics ($k = 0$) reduces to following expression

$$F_R = \frac{(\eta_i^2 - \eta_t^2)^2 c_{it}^2 + (\cos\theta_i^2 - \cos\theta_t^2)^2 n_{it}^2}{(c_{it}(\eta_i^2 + \eta_t^2) + n_{it}(\cos\theta_i^2 + \cos\theta_t^2))^2}, \quad (2.14)$$

where c_{it} is $\cos\theta_i\cos\theta_t$ and n_{it} is $\eta_i\eta_t$.

An important property of these equations is that they can be applied without regard to the direction of propagation [225]. We also remark that there is no absorption at an interface between dielectrics. Hence, the Fresnel coefficient for transmission, or transmissivity [175], denoted by F_T, can be obtained from F_R using a simple formula: $F_T = 1 - F_R$. Once light is transmitted into a medium, absorption may occur as discussed in the following sections.

2.3 LIGHT INTERACTIONS WITH MATTER

The problem of determining the appearance of an object or an environment involves the simulation of light interactions with matter, which involves three main processes: emission, scattering, and absorption. At the atomic level, these interactions occur between light (photons) and electrons, and they can be reduced to three basic actions: a photon goes from place to place, an electron goes from place to place, and an electron emits or absorbs a photon [86].

2.3.1 Emission

Man made and natural light sources emit light with characteristic spectral distributions, which depend on the nature of the emission process. The processes of light emission can be divided into two types: thermal and luminescent [13, 135].

Thermal emissions are due to the material radiating excess heat energy in the form of light (Figure 2.5). For these materials, called thermal radiators, the amount of light emitted is primarily dependent on the nature of the material and its temperature. For example, an incandescent light bulb is a thermal radiator where an electric current is run through a filament. The electrical

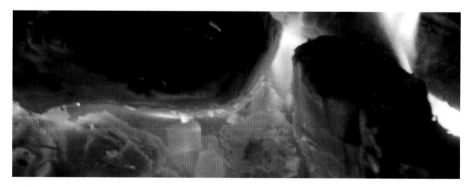

FIGURE 2.5

Photograph illustrating light emissions produced during a burning process. The yellow-orange glow corresponds to visible thermal emissions due to the high temperature of the burning wood pieces.

resistance of the filament causes its temperature to increase. As the temperature increases, energy is dispersed into the environment in the form of heat and light.

An ideal thermal radiator of uniform temperature whose radiant exitance in all parts of the spectrum is the maximum obtainable from any thermal radiator at the same temperature is called a *blackbody* [119]. Although no material reaches the theoretical maximum of a blackbody, it is sometimes convenient to describe the emissive properties of a material by specifying, on a wavelength-by-wavelength basis, the fraction of light it generates with respect to a blackbody [100]. For example, solar radiation arrives at the Earth's atmosphere with a spectral energy distribution similar to a blackbody radiator of $5800°K$ [231].

Luminescent emissions are due to energy arriving from elsewhere, which is stored in the material and emitted (after a short period of time) as photons. The incident energy, primarily due to factors other than temperature, causes the excitation of electrons of the material. These electrons in the outer and incomplete inner shells move to a higher energy state within the atom. When an electron returns to the ground state, a photon is emitted. The wavelength of the emitted photon will depend on the atomic structure of the material and the magnitude of the incoming energy. Typically, an electron remains in its excited state for about 10^{-9} s [275]. If there is a much longer delay and the electron emits a photon in the visible range, having being originally excited by a photon of differing (usually shorter) wavelength, the process is called phosphorescence. The distinction between phosphorescence and fluorescence is a matter of scale (time), with the latter usually taking less than 10^{-8} s [13].

A phosphor is defined as a luminescent material that absorbs energy and reemits it over some period of time, which is associated with the lifetime of the excited electron. Most phosphors are inorganic, i.e., carbon-free, crystals that contain structural and impurity defects. Some of these materials are used in TV screens and computer monitors (cathode ray tube [CRT]).

As described by Williamson and Cummins [275], atoms can be excited in many ways other than absorbing a photon. The term *phosphorescence* was originally applied to light given off by the reactive element phosphorous and chemically similar substances when left exposed to air. They spontaneously combine with oxygen in a slow reaction and in the process emit light. This process of light emission as a result of a chemical reaction is called chemiluminescence [275]. A related effect is bioluminescence, when light is produced by chemical reactions associated with biological activity. When one hard object is sharply stricken against another, we may observe a "spark" or light emission termed *triboluminescence*. Excitation is also possible due to the impact of high-energy particles, which may cause impressive light emissions such as those found in aurorae and space nebulae [24].

2.3.2 Scattering

The term *scattering* refers to the deflection of light through collisions with molecules, particles (an aggregation of sufficiently many molecules), or multiple particles. Besides the change in direction, the energy of the incident light may also be weakened (attenuated) in the process. The scattering processes can be divided into three groups: molecular scattering [168], particle scattering [36], and surface scattering [46]. While the study of the first two groups usually employs wave optics concepts, the study of the latter group usually employs ray optics concepts. The main types of scattering occurring in nature, namely Rayleigh scattering, Mie scattering, and reflective-refractive scattering, are associated with these three groups respectively.

Rayleigh, or molecular, scattering [242] occurs when the wavelength of the incident light is somewhat larger than the molecules, or particles. This type of scattering is proportional to the fourth power of the frequency, i.e., the shorter wavelengths are preferentially attenuated (Figure 2.6). Mie, or aerosol, scattering [168] occurs when the wavelength of the incident light is comparable to the size of the molecules, or particles. As the particles get larger, scattering tends to be more uniform across the light spectrum (Figure 2.6). For practical purposes, the error in applying Rayleigh, rather than Mie theory, to small particles is less than 1% when the radius of the particle is smaller or equal to 0.03λ [168].

FIGURE 2.6

Photograph depicting examples of Mie and Rayleigh scattering in nature. The smaller air molecules in the sky preferentially scatter light in the blue end of the spectrum (Rayleigh), while the suspended water particles in the cloud scatter light more uniformly across the spectrum (Mie).

FIGURE 2.7

Photograph depicting examples of reflective-refractive scattering.

Reflective-refractive, or geometrical optics, scattering occurs when the size of the particles is much larger than the wavelength of incident light (Figure 2.7). This type of scattering accounts for most of the internal scattering occurring in organic tissues such as plant leaves and human skin. It is mainly caused by the arrangement of tissues, and the refractive differences, which, for the most part, are associated with air-cell wall interfaces with respect to cells whose dimensions are quite large compared to the wavelength of light. Due to its dependency on refractive differences, the variations across the spectrum are directly associated with the wavelength dependency of the refractive indices of the materials.

2.3.3 Absorption

Once light is transmitted into a medium, it may be absorbed. In a dielectric, this may happen if there are absorptive elements, such as pigments and dyes, inside the medium (Figure 2.8). Pigments are materials that exhibit selective scattering and selective absorption, while dyes exhibit selective absorption and some luminescence produced due to excitation [84]. The resulting spectral distribution of light going through an absorption process depends on the absorption spectrum of the absorptive element. This spectrum, in turn, is determined by the type of chromophore, or molecular functional group, affecting the capture of photons. A chromophore is a region in a molecule characterized by an imbalance of charge (energy) between two different molecular orbitals. In such regions, electrons can be excited to higher states by only a relatively small amount of energy.

FIGURE 2.8

Photographs illustrating the absorption of light by organic (left) and inorganic (right) materials dispersed in the water.

When light with the right frequency (or wavelength) hits a chromophore, it is absorbed, i.e., it excites an electron from its ground state into an excited state [175]. If the incident light does not have the right frequency, it will not be absorbed, and it may be scattered instead. Many molecules found in living organisms have a number of features that help to reduce the energy needed to excite their electrons, and thus provide their color [84], which is determined by the spectral distribution of the absorbed light (Section 2.6). For instance, their atoms are often arranged in long chains and they may contain a transition metal. A transition metal is one of the group of elements (titanium, vanadium, chromium, manganese, iron, cobalt, nickel, and copper) that have unfilled electron orbits available for electron excitation. For example, iron atoms in hemoglobin, whose name derives from the combination of *heme* (iron) and *globin* (globular protein), are responsible for the red color of blood.

The absorption spectra of pigments, such as chlorophyll and hemoglobin, are usually given in terms of their specific absorption coefficient (s.a.c.), denoted by ζ. The s.a.c. can be obtained through direct measurements, or by dividing the material's molar extinction coefficient, ε, by the material's molecular weight. Its units depend on how the concentration of the material is measured. This issue is examined more closely in Section 2.6.1. In the case of inorganic materials (e.g., water and iron oxides), the absorption spectra is usually given in terms of their extinction coefficient. The s.a.c. of these materials can be obtained using the following formula [32] whose resulting values are provided in inverse length units:

$$\zeta(\lambda) = \frac{k(\lambda)4\pi}{\lambda}. \tag{2.15}$$

The extinction coefficient of certain inorganic materials like pure water is fairly low across the visible portion of the light spectrum (Figure 2.3).

FIGURE 2.9

Photograph illustrating a calved bergy bit (small iceberg) in Alaska. Its deep blue color is due to the extinction of light in the red portion of the visible spectrum as it travels through the heavily packed ice layers. Its surface, on the other hand, appears white due to the scattering of light caused by the presence of material irregularities such as tiny air bubbles [35].

For this reason, the absorption of visible light by these materials is usually assumed to be negligible. We note, however, that seemingly small extinction coefficient differences may have a significant effect on the appearance of a material under certain conditions (Figure 2.9).

The absorption spectra of a given pigment present in a tissue is usually obtained under in-vitro conditions, i.e., the pigment is extracted from the tissue using various techniques (e.g., employing organic solvents in the case of chlorophyll). Due to differences in the surrounding environment, distribution and state of these pigments, their absorption spectra under in-vitro conditions differ from that of pigments under in-vivo conditions [81]. For example, in-vivo pigments are usually inhomogeneously distributed in the tissues. Consequently, light transport models that use in-vitro data to characterize pigments should account for these changes to improve the accuracy of their results. One alternative to take them into account is to adjust the in-vitro absorption spectra of pigments according to the lengthening of the optical path in this tissue, also called ratio of intensification [40] or factor of intensification [213] in plant sciences. In biomedical applications, a similar adjustment is performed using a quantity called differential path length [60]. The factor of intensification, denoted by B, represents a combination of light that passes through tissues without encountering an absorber (sieve effect) and the light that is scattered and has an increased path length (detour effect). These two phenomena have opposite outcomes: the sieve effect lowers absorption (especially at or near wavelengths for which the absorption has a maximum value), whereas the detour effect increases absorption (especially at or near wavelengths for which the absorption has a minimum value) [81]. In dispersive samples, the absorption is usually enhanced by the combination of these two effects [18]. The main difficulty in employing this correction factor is the scarcity of available measured data to allow its accurate quantification for different in-vivo conditions.

2.4 RADIOMETRIC QUANTITIES

The measurement of quantities associated with the transport of radiant energy is the object of detailed studies in the field of radiometry [77]. In this field, one can find a system of concepts, terminology, mathematical relationships, measurement instruments, and units of measure devised to not only describe and measure radiation but also its interactions with matter [169]. We should note that radiation (light) can also be measured using a similar system provided in the field of photometry, which is intended to account for light detection by the human eye.

In this section, we define the radiometric quantities usually employed in light transport simulations, namely radiant power, radiant intensity, radiance, and radiant exitance. These quantities describe measurements of light integrated over all wavelengths. The adjective "spectral" is used to characterize the same measurements evaluated at a specific wavelength λ. For the sake of simplicity, we will, however, omit this adjective in the following definitions and use the same terms for wavelength dependent and wavelength integrated quantities.

Radiant energy, denoted by Q (measured in joules, J), is a fundamental quantity representing the energy of a packet of rays. In light transport simulations, it is usually assumed that there is a steady state of energy flow representing the amount of light hitting a surface or film plane during a set period of time. This quantity, measured as radiant power or flux, denoted by Φ (measured in Watts, W, or Js^{-1}), is therefore often used in these simulations.

The amount of radiant power traveling from a source in a certain direction, per unit of solid angle, is called the radiant intensity and denoted by I (measured in Wsr^{-1}). The concept of solid angle can be described as the three-dimensional analog to the two-dimensional concept of angle [100]. For example, the solid angle, denoted by ω, subtended by an area A on a sphere with radius r is equal to Ar^{-2}. This quantity is the measure of the angle in steradians (radians squared or simply sr). Assuming a source located in a point x, a differential distribution of directions can be represented by a differential area, denoted by dA, on a sphere with center in x and radius r equal to one. Considering the representation of points on the sphere using spherical coordinates (polar and azimuthal angles θ and ϕ, respectively), such a differential area is given by d$A = r^2 \sin\theta\,\mathrm{d}\theta\,\mathrm{d}\phi$. This results in a differential solid angle given by d$\omega = \sin\theta\,\mathrm{d}\theta\,\mathrm{d}\phi$ [52]. Integrating over the entire sphere, one obtains the total solid angle, which corresponds to 4π.

Finally, radiance, denoted by L (measured in Wsr^{-1}m^{-2}), is neither dependent on the size of the object being viewed nor on the distance to the viewer. It can be written in terms of radiant power, radiant intensity, or radiant exitance, M (measured in Wm^{-2}). More specifically, the radiance at a point x of a surface and in a direction ψ (usually represented by a pair of spherical coordinates θ and ϕ) can be expressed as

$$L(x,\psi) = \frac{\mathrm{d}I(x,\psi)}{\mathrm{d}A\cos\theta} = \frac{\mathrm{d}^2\Phi(x,\psi)}{\mathrm{d}\vec{\omega}\,\mathrm{d}A\cos\theta} = \frac{\mathrm{d}M(x,\psi)}{\mathrm{d}\vec{\omega}\cos\theta}. \qquad (2.16)$$

2.5 TISSUE OPTICS DEFINITIONS AND TERMINOLOGY

A level of abstraction commonly employed in tissue optics simulations corresponds to the characterization of organic tissues as a random turbid media with volumetric scattering and absorption properties [201]. These properties are represented by the volumetric absorption and scattering coefficients. Since the term volumetric is usually omitted in the tissue optics literature, for the sake of consistency, we will also omit the term volumetric throughout this book.

The absorption coefficient is obtained by multiplying the absorption cross section of an absorber by the density of the absorbers in the medium [123]:

$$\mu_a(\lambda) = C_a(\lambda)\vartheta_a, \qquad (2.17)$$

where $C_a(\lambda)$ is the absorption cross section and ϑ_a is the density of the absorbers in the medium.

The absorption cross section of a particle corresponds to the total power absorbed by the particle, and it is a function of the orientation of the particle and the state of polarization of the incident light [262].

Similarly, the scattering coefficient is obtained by multiplying the scattering cross section of a scatterer particle by the density of the scatterers in the medium [123]:

$$\mu_s(\lambda) = C_s(\lambda)\vartheta_s, \qquad (2.18)$$

where $C_s(\lambda)$ is the scattering cross section and ϑ_s is the density of the scatterers in the medium.

The scattering cross section of a particle corresponds to the total power scattered by the particle, and it is also a function of the orientation of the particle and the state of polarization of the incident light [262].

The sum of scattering and absorption coefficients gives the attenuation coefficient μ. These coefficients are typically measured in inverse length units, while the cross sections are given in area units and the densities are given in inverse volume units. In tissue optics, these coefficients are usually used to describe the optical properties of whole tissues instead of their basic constituents. This approach may introduce undue inaccuracies in simulations of light propagation in organic tissues since, due to scarcity of direct measured data, these coefficients are usually determined using model inversion procedures (Chapter 5).

Three other parameters are commonly used to simulate light propagation in a tissue assumed to be a random medium with volumetric properties: albedo, optical depth, and phase function. The albedo is a dimensionless parameter defined as the ratio between the scattering coefficient and the attenuation coefficient:

$$\gamma(\lambda) = \frac{\mu_s(\lambda)}{\mu(\lambda)}. \tag{2.19}$$

The optical depth represents the product of the tissue thickness, denoted by h, and the attenuation coefficient [201]:

$$\varrho(\lambda) = h\mu(\lambda). \tag{2.20}$$

An optical depth of 1 indicates that there is a probability of approximately 36.78% that light will traverse the tissue without being neither scattered nor absorbed [201], i.e., the ratio between the incident and the transmitted power corresponds to

$$\frac{\Phi_t(\lambda)}{\Phi_i(\lambda)} = e^{-\varrho(\lambda)} = e^{-1} = 0.3678.$$

When light hits a particle with an index of refraction different from that of its environment, the light is scattered. The direction of scattering is characterized by the polar angle α at which the light is bent and an azimuthal angle β in a plane normal to the direction of incidence (Figure 2.10). A phase function, denoted by $\Gamma(\alpha, \beta)$, describes the amount of light scattered from the

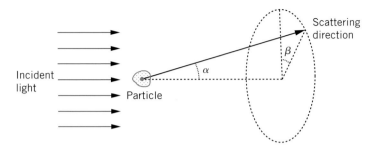

FIGURE 2.10

Sketch describing a scattering direction represented by a polar angle α and an azimuthal angle β.

direction of incidence to the direction of scattering [262], i.e., it represents a single scattering event. The probability of light scattering through an angle α after n scattering events is given by a multiple-scattered phase function, a concept used by Tessendorf and Wasdon [246] to simulate multiple scattering in clouds.

In astrophysics, a phase function is treated as a probability distribution, and its normalization requires that the integral over all angles to be equal to one:

$$\int_{4\pi} \Gamma(\alpha, \beta) \, d\omega = 1. \tag{2.21}$$

The average cosine of the phase function, or asymmetry factor, is used to describe the degree of asymmetry of the phase function. It can be defined as

$$g = \int_{4\pi} \Gamma(\alpha, \beta) \cos\alpha \, d\omega. \tag{2.22}$$

The simplest phase function is the symmetric one ($g = 0$):

$$\Gamma(\alpha, \beta) = \frac{1}{4\pi}. \tag{2.23}$$

Another phase function often used in tissue optics simulations is the Rayleigh phase function [168] associated with Rayleigh scattering. In this case, the function is assumed to be symmetric with respect to the azimuthal angle β, and the direction changes with respect to the polar angle α are given by

$$\Gamma_R(\alpha) = \frac{3}{4}(1 + \cos^2\alpha). \tag{2.24}$$

The name "phase function" has no relation to the phase of the electromagnetic wave (light). It has its origins in astronomy, where it refers to lunar phases [123]. Coincidentally, one of the most commonly used phase functions in tissue optics, namely the Henyey–Greenstein phase function (HGPF), was originally formulated for astrophysical applications [113].

The HGPF was proposed by Henyey and Greenstein [113] to approximate Mie scattering in their study of diffuse radiation in galaxies (Figure 2.11). In other words, the HGPF is not based on the mechanistic theory of scattering [128]. The formula proposed by Henyey and Greenstein is given by

$$\Gamma_{HG}(g,\alpha) = \frac{1}{4\pi} \frac{1-g^2}{(1+g^2-2g\cos\alpha)^{\frac{3}{2}}}. \tag{2.25}$$

The HGPF is actually a function of three parameters: g, α, and β. It just happens that an azimuthal symmetry is also assumed in this case. By varying the parameter g in the range $-1 \leq g \leq 1$, it is possible to characterize HGPFs ranging from a completely backward-throwing ($g < 0$) to a completely forward-throwing ($g > 0$) form (Figure 2.12).

The HGPF as defined in Equation 2.25 cannot, however, be used to describe simultaneous forward and backward lobes, which are typical in many

FIGURE 2.11

Photograph depicting the Andromeda galaxy.

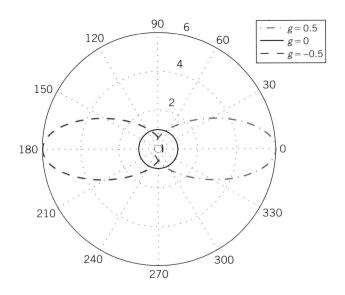

FIGURE 2.12

Scattering diagrams illustrating different scattering profiles provided by the HGPF. By varying the parameter g, it is possible to obtain a backward lobe ($g < 0$) or a forward lobe ($g > 0$).

scattering situations [277]. For this reason, astrophysicists proposed variations based on the superposition of two HGPFs [138, 257, 277]:

$$\Gamma(g_1, g_2, \alpha, u) = u\,\Gamma_{\text{HG}}(g_1, \alpha) + (1 - u)\Gamma_{\text{HG}}(g_2, \alpha), \qquad (2.26)$$

where u is the phase function weight defined in the interval $[0, 1]$.

Figure 2.13 illustrates three scattering profiles that can be provided by the two-term HGPF.

The asymmetry factor g is oftentimes called anisotropy factor. Throughout this book, we will employ the term *asymmetry* when referring to g since the use of the term *anisotropy* in this context may result in an inappropriate association with the macroscopic anisotropic behavior of a given material. Such a behavior, or anisotropy, is observed when the material is rotated around its normal, while the light and the viewer directions remain unchanged, and the light intensity reflected to the viewer varies. Most natural materials are anisotropic, i.e., their reflection profile depends on both the polar and the azimuthal angles measured from the material's normal, and used to define the direction of incidence of the incoming light. For practical reasons (Chapter 10), however, the dependence of their reflection profile on the azimuthal angle of light incidence is often omitted in light transport simulations leading to the modeling of their appearance.

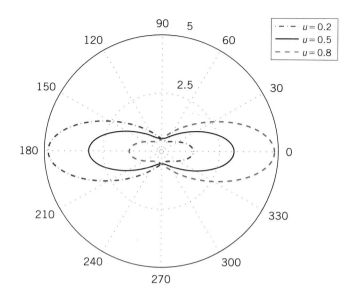

FIGURE 2.13

Scattering diagrams illustrating different scattering profiles provided by the superposition of the two-term HGPF with different asymmetry factors ($g_1 = 0.5$ and $g_2 = -0.5$) and varying their respective weights (parameter u).

2.6 MEASUREMENT OF APPEARANCE

The group of measurements necessary to characterize both the color and surface finish of a material is called its *measurement of appearance* [119]. These measurements involve the spectral and the spatial energy distribution of propagated light. The following definitions closely follow those provided in the seminal book by Hunter and Harold [119] on this topic.

The variations in the spectral distribution of the propagated light affect appearance characteristics such as hue, lightness, and saturation (Figure 2.14). Hue is the attribute of color perception by means of which an object is judged to be red, yellow, green, blue, purple, and so forth (Section 3.4). Lightness is the attribute by which white objects are distinguished from gray objects and light from dark-colored objects. Finally, saturation is the attribute that indicates the degree of departure from the gray of the same lightness.

The changes in the spatial distribution of the propagated light affect appearance characteristics such as glossiness, reflection haze, transmission haze, luster, and translucency (Figure 2.15). The reflection haze corresponds to the scattering of reflected light in directions near that of specular reflection by a specimen having a glossy surface. The transmission haze corresponds to

FIGURE 2.14

Photograph illustrating different spectral distributions of red and green maple leaves.

FIGURE 2.15

Photograph illustrating appearance characteristics, such as glossiness and translucency, determined by different spatial distributions of light.

the scattering of transmitted light within or at the surface of a nearly clear specimen [119]. The luster, or contrast gloss, corresponds to the gloss associated with contrasts of bright and less bright adjacent areas of the surface of an object. Luster increases with increased ratio between light reflected in the specular direction and that reflected in those diffuse directions, which are adjacent to the specular direction. Finally, the translucency property of a material corresponds to incoherent transmission, i.e., a significant portion of the transmitted light undergoes scattering [119].

2.6.1 Measuring the spectral distribution of light

The spectral energy distribution of the propagated light is usually measured in terms of reflectance and transmittance. There are nine different representations of reflectance and transmittance. These representations depend on the incident and propagated (collected) light geometries, which are designated as directional, conical, and hemispherical. For the sake of consistency with the literature on this topic, the following descriptions of these geometries closely follow those provided by Nicodemus et al. [181]. The directional geometry designates a differential solid angle $d\omega$ about a single direction ψ (represented by a pair of spherical coordinates θ and ϕ). The conical geometry designates a solid angle ω of any configuration (e.g., a right circular cone). Finally, a hemispherical geometry designates a full hemispherical solid angle $\omega = 2\pi$. Unless otherwise stated, when we refer to reflectance and transmittance in this book, we will be assuming a directional-hemispherical geometry. Furthermore, for the sake of representation simplicity, the reflectance and transmittance dependence on the point of incidence is going to be omitted in the expressions presented in this section.

Reflectance corresponds to the fraction of light at wavelength λ incident from a direction ψ_i at a point x that is neither absorbed into nor transmitted through a given surface, and it is denoted by $\rho(\psi_i, \lambda)$. Alternatively, the reflectance can be defined as the spectral power distribution of the reflected light, i.e., the ratio of the reflected flux, Φ_r, to the incident flux, Φ_i:

$$\rho(\psi_i, \lambda) = \frac{\Phi_r(\lambda)}{\Phi_i(\lambda)}. \tag{2.27}$$

Similarly, the fraction of light transmitted through the surface is called the transmittance, denoted by $\tau(\lambda)$. It represents the spectral power distribution of the transmitted light, i.e., the ratio of the transmitted flux, Φ_t, to the incident flux:

$$\tau(\psi_i, \lambda) = \frac{\Phi_t(\lambda)}{\Phi_i(\lambda)}. \tag{2.28}$$

The light that is neither reflected nor transmitted by the materials is absorbed. The parameter that describes the amount of absorbed light is absorptance, denoted by $\mathcal{A}(\lambda)$ [13]. Due to energy conservation, the sum of reflectance, transmittance, and absorptance is equal to one.

Another radiometric quantity employed in the measurement of spectral distribution of light is the reflectance factor, denoted by $R(\psi_i, \lambda)$ [279].

It represents the ratio of the reflected flux from a surface to the flux that would have been reflected by a perfectly diffuse surface, Φ_{pd}, in the same circumstances:

$$R(\psi_i, \lambda) = \frac{\Phi_r(\lambda)}{\Phi_{pd}(\lambda)}.$$

(2.29)

The transmittance of a homogeneous material, after correction for surface losses, varies in accordance with Lambert's law of absorption (Figure 2.16), also called Bouguer's law [163]. This law states that the loss due to the process of absorption is proportional to the thickness of the medium (or the distance traveled by the light in the material) and to a constant of proportionality represented by its specific absorption coefficient. Assuming a direction of incidence perpendicular to the medium, this law is usually expressed as follows [175]:

$$\tau(\lambda) = \frac{\Phi_t(\lambda)}{\Phi_i(\lambda)} = e^{-\zeta(\lambda)h}.$$

(2.30)

Similarly, the Beer's law [159] states that for a dye solution, the loss due to the process of absorption is proportional to the dye concentration. Combining Beer's law with Lambert's law [163] for samples of thickness h and concentration c results in the following expression for the transmittance of a homogeneous material, also called Beer–Lambert law:

$$\tau(\lambda) = e^{-\zeta(\lambda)ch}.$$

(2.31)

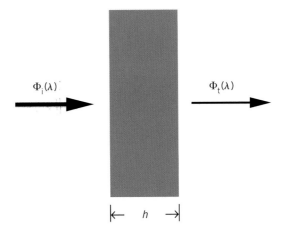

FIGURE 2.16

Loss of light at wavelength λ in a medium of thickness h.

The Beer–Lambert law can be generalized in different ways. For example, if the sample has n different pigments with concentrations c_i (where $i = 1, \ldots, n$), this law becomes

$$\tau(\lambda) = e^{-\left(\sum_{i=1}^{n} \zeta_i(\lambda) c_i\right)h}. \tag{2.32}$$

Furthermore, assuming the availability of factors of intensification $B_i(\lambda)$ (where $i = 1, \ldots, n$) for each pigment present in the sample, the modified Beer–Lambert law [60] can be applied:

$$\tau(\lambda) = e^{-\left(\sum_{i=1}^{n} \zeta_i(\lambda) B_i(\lambda) c_i\right)h}. \tag{2.33}$$

Finally, considering a polar propagation angle θ different from zero [10], the modified Beer–Lambert law becomes

$$\tau(\lambda) = e^{-\left(\sum_{i=1}^{n} \zeta_i(\lambda) B_i c_i\right)h \sec\theta}. \tag{2.34}$$

2.6.2 Measuring the spatial distribution of light

Similar to the previous section, the following descriptions of radiometric quantities used to represent spatial patterns of light distribution closely follow those provided by Nicodemus et al. [181]. Accordingly, such patterns are represented by the bidirectional scattering-surface distribution function (BSSDF) or its components: the bidirectional scattering-surface reflectance distribution function (BSSRDF) and its counterpart the bidirectional scattering-surface transmittance distribution function (BSSTDF).

The BSSDF is considered to be a difficult function to measure, store, and compute due to its dependency on four parameters: the incidence and outgoing directions, the wavelength, and the position on the surface (Figure 2.17). For this reason, sometimes it is more practical to make simplifying assumptions about the material in order to obtain a more tractable function and still obtain a useful degree of approximation for the cases of interest [181]. For example, if one assumes that the scattering properties of a given material are uniform, the dependence on the location of the point of observation (reflection) can be omitted [181]. In this case, one can work with a simpler function, namely the bidirectional scattering distribution function (BSDF or simply BDF), which can also be decomposed into two components: the bidirectional reflectance distribution function (BRDF) and the bidirectional transmittance distribution function (BTDF).

The BDF, denoted by f, can be expressed in terms of the ratio between the radiance propagated at a surface in the direction ψ and the radiant energy

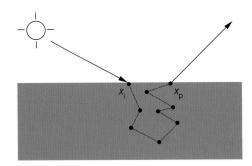

FIGURE 2.17

Sketch depicting the dependence of the BSSDF on the point of propagation (or observation), denoted by x_p, which is usually not coincident with the point of incidence, denoted by x_i.

(per unit of area and per unit of time) incident from a direction ψ_i at the surface [181]:

$$f(\psi_i, \psi, \lambda) = \frac{dL(\psi, \lambda)}{L_i(\psi_i, \lambda) d\omega_i \cos\theta_i}, \tag{2.35}$$

where $dL(\psi, \lambda)$ is the radiance propagated in a direction ψ and $L_i(\psi_i, \lambda)$ is the incident radiance in a direction ψ_i.

The BRDF component, denoted by f_r, is obtained by considering the reflection direction in Equation 2.35. Similarly, the BTDF component, f_t, is obtained by considering the transmission direction in Equation 2.35.

An important property of the BDF is its symmetry or reciprocity condition, which is based on the *Helmholtz reciprocity rule* [51, 268]. This condition states that the BDF for a particular point and incoming and outgoing directions remains the same if these directions are exchanged. Quantitatively, this condition can be expressed as

$$f(\psi_i, \psi, \lambda) = f(\psi, \psi_i, \lambda). \tag{2.36}$$

Another important property of the BDFs is that they must be normalized, i.e., conserve energy. This means that the total energy propagated in response to some irradiance must be no more than the energy received [100]. In other words, for any incoming direction ψ_i, the radiant power propagated over the hemisphere, denoted by Ω, can never be more than the incident radiant power [151]. Any radiant power that is not propagated is absorbed. Formally, in the case of reflection of light, the directional-hemispherical reflectance should therefore be less than or at most equal to one [13]:

$$\rho(\psi_i, 2\pi, \lambda) = \int_{\Omega\psi} f_r(\psi_i, \psi, \lambda) \cos\theta \, d\vec{\omega} \leq 1, \quad \forall \psi_i. \tag{2.37}$$

A similar relation given in terms of the directional-hemispherical transmittance and the BTDF is used for the transmission of light.

Sometimes, when energy transport or energy balance is of concern as opposed to lighting at a point, it is more convenient to work with the radiant power (radiant flux) than with the radiance [225]. Under these circumstances, it is more natural to describe the surface reflection and transmission properties in terms of the *scattering probability function* (SPF), denoted by s. This quantity describes the amount of energy scattered at a surface in a direction ψ and at a wavelength λ as

$$s(\psi_i, \psi, \lambda) = \frac{dI(\psi, \lambda)}{\rho(\psi_i, \lambda) d\Phi(\psi_i, \lambda)}. \tag{2.38}$$

The term $\rho(\psi_i, \lambda)$ appears in the numerator of Equation 2.38 when we are dealing with reflection of light. In the case of transmission, an expression similar to Equation 2.38 is used, in which $\rho(\psi_i, \lambda)$ is replaced by $\tau(\psi_i, \lambda)$.

Image-synthesis context

Different image-synthesis frameworks can be used to realistically render the appearance of human skin. They share, however, a similar pipeline structure (Figure 3.1). The stages of this pipeline include, but are not limited to, the geometrical representation of skin surfaces, the spectral sampling of light sources, the modeling of (local) light and skin interactions, the computation of environmental (global) light transfers, and the conversion of the resulting radiometric quantities into appearance attributes (e.g., color and glossiness).

In this chapter, we provide an overview on how the skin-lighting models examined in this book fit in these frameworks. We start by presenting a concise review of key-rendering concepts and techniques used to simulate global light transport. We remark that, although this presentation is aimed at computer graphics applications, these concepts and techniques are also relevant for applications in several other fields such as biomedical optics, remote sensing, and computer vision. At the local (tissue) level, deterministic and nondeterministic approaches can be used in the modeling of light and skin interactions. Although a comprehensive review of these approaches is beyond the scope of this book, we examine the main characteristics of methods representative of each group.

We also provide a concise description of procedures that can be used in the evaluation of the predictive capabilities of skin-lighting models. These procedures not only contribute to the identification of possible error sources affecting the final rendering results, but they can also be used offline to derive data from previously validated models. Such precomputed data, in turn, can be incorporated in simulations of light transport in skin tissues performed on the fly, without incurring excessive computational costs.

The last stage of the image-synthesis pipeline involves mapping the resulting spectral signals (e.g., given in terms of radiances) to display values. This

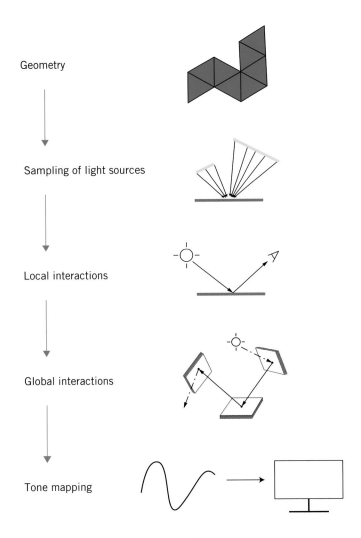

Geometry

Sampling of light sources

Local interactions

Global interactions

Tone mapping

FIGURE 3.1

Diagram illustrating the main stages of an image-synthesis pipeline.

is an important but, often, underappreciated task. It must account for the physical characteristics of the display device and the perceptual characteristics of the viewer. For example, incident light interacts with a material, and it is reflected back to the environment (Figure 3.2). Our perception of the color of the material depends on how the photoreceptors in our eyes respond to this propagated signal. In computer graphics applications, we are interested in mapping this spectral information into a given color system. This wavelength-dependent color-specification process has been extensively examined in the colorimetry [118, 188] and computer graphics literatures

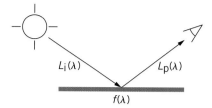

FIGURE 3.2

Sketch illustrating the basic steps of a light-propagation process leading to the perception of appearance attributes of a material. Incident light, represented in terms of spectral incident radiance $L_i(\lambda)$, interacts with a material characterized by a BDF $f(\lambda)$. The resulting propagated light, represented in terms of spectral propagated radiance $L_p(\lambda)$, eventually reaches our visual system, where it is translated to appearance attributes such as color.

[29, 240]. Nonetheless, because color is a major component of skin appearance, we close this chapter with an outline of the basic steps involved in the conversion of spectral radiometric quantities to color values.

3.1 GLOBAL LIGHT TRANSPORT

In order to render the image of a scene or object, we need to compute the radiance values at selected points. This process involves the solution of an integral equation describing light transfer in the environment of interest. This integral light transport equation, also known as the rendering equation in computer graphics [136], can be expressed in different forms, depending on the application field. For consistency with the graphics literature, we will express it in terms of radiances on the basis of the ray law (the radiance is constant along a line of sight between objects [225]), and the definition of the bidirectional scattering distribution function (BDF) (Section 2.6.2). In a simplified form, it is given by

$$\underbrace{L(x,\psi,\lambda)}_{\text{total}} = \underbrace{L_e(x,\psi,\lambda)}_{\text{emitted}} + \underbrace{L_p(x,\psi,\lambda)}_{\text{propagated}} \qquad (3.1)$$

Equation 3.1 states that the radiance of a point x on a surface, in a direction ψ and at wavelength λ is given by the sum of the emitted radiance component, L_e, and the propagated radiance component, L_p. Usually L_e is known from the input data, and the computation of L_p constitutes the major computational problem. The term L_p can be written as an integral over all the surfaces within the environment. Accordingly, using the geometry described

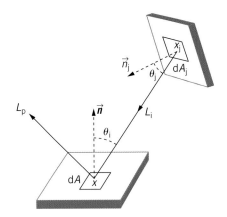

FIGURE 3.3

Geometry for computing the propagated radiance L_p at a point x as an integral over all the surfaces within the environment of interest. The differential areas surrounding points x and x_j are represented by dA and dA_j, respectively.

in Figure 3.3, this term can be represented by

$$L_p(x, \psi, \lambda) = \int_{\text{all } x_j} f(x, \psi, \psi_i, \lambda) L_i(x, \psi_i, \lambda) \cos\theta_i V(x, x_j) \frac{\cos\theta_j dA_j}{\|x_j - x\|^2}, \quad (3.2)$$

where θ_j is the angle between the normal at x_j and the direction of incidence, dA_j is the differential area surrounding x_j, and V is the visibility term.

The direction of incidence, represented by ψ_i, corresponds to the vector given by $x_j - x$, and the visibility term $V(x, x_j)$ is equal to one if a point x_j on a surface can "see" a point x on the other surface, and zero otherwise. Equation 3.2 is commonly used by deterministic rendering methods such as those based on the classic radiosity approach [52].

Alternatively, L_p can also be expressed in terms of all directions visible to a point x (Figure 3.4). This representation of L_p is suitable for nondeterministic rendering methods based on stochastic ray-tracing techniques [225], and it is given by

$$L_p(x, \psi, \lambda) = \int_{\text{incoming } \psi_i} f(x, \psi, \psi_i, \lambda) L_i(x, \psi_i, \lambda) \cos\theta_i d\vec{\omega}_i, \quad (3.3)$$

where θ_i is the angle between the normal at x and the direction of incidence and $d\vec{\omega}_i$ is the differential solid angle where L_i arrives.

Several global-illumination methods [52, 75, 268] have been proposed to the solution of Equation 3.1, and no single method is superior in all

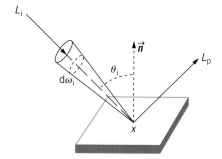

FIGURE 3.4

Geometry for computing the propagated radiance L_p at a point x in terms of all directions visible to this point. The direction in which the incident radiance L_i arrives is represented by the differential solid angle $d\omega_i$.

cases. In order to illustrate the process of global light transport simulation, we choose to expand on the stochastic ray-tracing approach. This pragmatic choice is motivated by the fact that such an approach is often used in both levels of light transport simulation, global and local, which simplifies the discussion of issues related to their interplay. Before getting to this approach, however, we should define some relevant Monte Carlo concepts [108].

3.1.1 Monte Carlo concepts

The classic Monte Carlo method was originally proposed by Metropolis and Ulam [174] as a statistical approach to the solution of integro-differential equations that occur in various branches of natural sciences, including light transport simulation. Accordingly, the integral term of Equation 3.3 can be efficiently estimated using Monte Carlo techniques such as *importance sampling* [108]. The idea behind this technique is simple. If the integrand is a product of two functions, and we know one of them, we can use this information to guide the sampling strategy used to solve the integral. For example, suppose that we need to determine the expected value, denoted by Υ, that results from the following integral involving a real-valued function q:

$$\Upsilon = \int q(x) \, dx. \qquad (3.4)$$

This integral can be rewritten as

$$\Upsilon = \int \frac{q(x)}{P(x)} P(x) \, dx, \qquad (3.5)$$

where $P(x)$ represents the importance function, also called the probability density function (PDF) [108].

The PDF needs to satisfy the following conditions [137, 151]:

- $P(x) \geq 0$ for each $x \in [0,1]$ for which $q(x) \neq 0$,

- $\int_0^1 P(x)\, dx = 1$,

- $\frac{q(x)}{P(x)} < \infty$ except perhaps on a (countable) set of points.

A technique called warping [227] is often used to solve Equation 3.5. It consists of generating uniform distributed random samples, denoted by ξ, in a canonical space [0,1]. These samples are then transformed so that their distribution matches the desired density given by P.

The key aspect of any importance sampling application is the selection of the PDF. For example, for the problem represented by Equation 3.5, an optimal PDF would be given by $P(x) = \kappa q(x)$, with the constant κ equal to $1/\Upsilon$. Clearly, this is not an option because if we already knew Υ, we would not need to use Monte Carlo techniques to estimate it. The practical solution is to choose a function $\tilde{P}(x)$ "close" to $P(x)$. We will examine this issue in more detail in the next section.

3.1.2 Path tracing overview

The series of scattered or absorbed states that can be assumed by a particle (in our case, a photon) is referred to as a random walk [108]. If the walk terminates after a number of steps, then the successive states are connected by the transition probabilities associated with scattering events and the termination probabilities associated with absorption events. The type of random walk in which the current state depends only on the previous state is known as a Markov chain [108].

The solution of the light transport (rendering) equation can be estimated by following a photon (or a ray) as it bounces through the environment. Each bounce is determined by sampling the integrand of Equation 3.3, using the current state of the photon (wavelength, position, and direction of propagation) as the integrand's parameters. This process is called path (history) tracing [100], and it was introduced to computer graphics by Kajiya [136]. It can be defined as a Markov chain in which a ray represents a photon or an ensemble of photons, depending on the power associated with the ray.

For the sake of simplicity, because the transmission is usually handled very similarly to reflection, we will focus on the bidirectional reflectance distribution function (BRDF) (f_r) in the following presentation.

To solve Equation 3.1 using importance sampling within a stochastic (Monte Carlo based) ray-tracing framework, new scattering directions have to be sampled, recursively, at each bounce location, represented by x, such that the reflected radiance is given by

$$L_r(x, \psi) = L_e(x, \psi)$$
$$+ f_r(x, \psi, \psi_i) L_e(x', \psi') \qquad (3.6)$$
$$+ f_r(x, \psi, \psi_i) f_r(x', \psi', \psi_i') L_e(x'', \psi'') + \cdots$$

In a path-tracing implementation of Equation 3.6, for example, the point x^n is chosen by sending a ray from x^{n-1} in the direction ψ^{n-1} represented by $x^n - x^{n-1}$ (Figure 3.5). This direction, in turn, is chosen according to a PDF (or directional probability density [227]) based on the reflection behavior of the materials (Figure 3.6), which can be quantitatively represented by their SPF (Section 2.6.2).

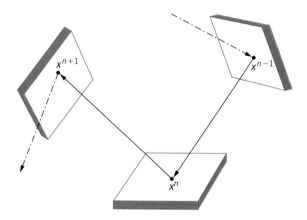

FIGURE 3.5

Sketch illustrating an example of a random walk observed in a path-tracing implementation.

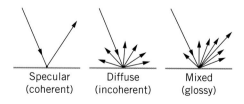

FIGURE 3.6

Examples of different reflection behaviors (reflected ray distributions).

Ideally, to choose reflected ray directions, one should be able to sample according to the following PDF given by

$$P(\psi) = \frac{f_r(x, \psi_i, \psi) \cos\theta_i}{\int_\Omega f_r(x, \psi_i, \psi) \cos\theta_i d\omega_i}. \qquad (3.7)$$

We remark that, in practice, one often has to use an approximating PDF. For example, usually diffuse reflection directions (Figure 3.6) are sampled using a PDF based on the reflection behavior of Lambertian materials, which follows a cosine distribution [181]. This PDF is given by

$$P_d(\alpha_d, \beta_d) = \frac{1}{\pi} \cos\alpha_d, \qquad (3.8)$$

where α_d and β_d represent the corresponding polar and azimuthal scattering angles.

Similarly, mixed or glossy reflection directions (Figure 3.6) are usually sampled using PDFs based on exponentiated cosine distributions. In these distributions [122, 153], a glossiness exponent, denoted by n_g, indicates how far ($n_g \gg 1$) the reflection behavior of the material is from a perfect Lambertian behavior. An example of such a PDF is given by

$$P_m(\alpha_m, \beta_m) = \frac{n_g + 1}{2\pi} \cos^{n_g}\alpha_m, \qquad (3.9)$$

where α_m and β_m represent the corresponding polar- and azimuthal-scattering angles, respectively.

The PDFs presented in Equations 3.8 and 3.9 can be sampled by selecting two stochastic variables, ξ_1 and ξ_2, representing random numbers uniformly distributed in the interval [0,1], and transforming them using the warping technique as follows.

Recall that if a random variable ξ ranges over a region $\mathfrak{R}$, then the probability that ξ will take on a value in some subregion $\mathfrak{R}_i \subset \mathfrak{R}$ is given by

$$\mathcal{P}(\xi \in \mathfrak{R}_i) = \int_{\xi' \in \mathfrak{R}_i} P(\xi') d\zeta(\xi') \quad (P : \mathfrak{R} \to \mathfrak{R}^1), \qquad (3.10)$$

where $\mathcal{P}(event)$, also called cumulative distribution function [108], is the probability that the *event* is true [225]. In computer graphics applications where the propagated radiances are expressed in terms of all directions visible to a point x (Equation 3.3), $d\zeta$ is represented by a differential solid angle.

Considering the PDF given by Equation 3.8, the corresponding cumulative distribution function is given by

$$\mathcal{P}(\alpha_d, \beta_d) = \int_0^\beta \int_0^\alpha \frac{\cos\alpha'}{\pi} \sin\alpha' \, d\alpha' \, d\beta'. \qquad (3.11)$$

The PDF in the integrand of Equation 3.11 is separable, and derivation techniques can be applied on each dimension to find the warping function used to generated the corresponding scattered directions [23]. This warping function is given by

$$(\alpha_d, \beta_d) = \left(\arccos\left(\sqrt{(1 - \xi_1)}\right), 2\pi\xi_2 \right), \qquad (3.12)$$

and the resulting polar (α_d) and azimuthal (β_d) angles can then be used to perturb the normal vector $\vec{n}$ (or $-\vec{n}$ in case of transmission) of the material in order to obtain a diffusively scattered ray.

Similarly, considering the PDF given by Equation 3.9, the corresponding cumulative distribution function is given by

$$\mathcal{P}(\alpha_m, \beta_m) = \int_0^\beta \int_0^\alpha \frac{n_g + 1}{2\pi} \cos^{n_g}\alpha' \sin\alpha' \, d\alpha' \, d\beta'. \qquad (3.13)$$

Because the PDF in the integrand of Equation 3.13 is also separable [225], derivation techniques can be applied on each dimension [23] to find the following warping function:

$$(\alpha_m, \beta_m) = (\arccos(1 - \xi_1)^{\frac{1}{n_g+1}}, 2\pi\xi_2), \qquad (3.14)$$

and the resulting polar (α_m) and azimuthal (β_m) angles can then be used to perturb the propagated reflect ray $\vec{r}$ (or $\vec{t}$ in case of transmission).

3.2 LOCAL LIGHT TRANSPORT

The global light transport simulation approaches outlined in the previous section rely on the accurate modeling of local interactions between light and matter. In fact, such an interplay between global and local light transport, as appropriately highlighted by Fournier [95], has several theoretical and practical implications such as the possibility of using similar equations and algorithms to simulate light transfers.

The methods commonly used to simulate local light transport can be divided into three groups: deterministic, nondeterministic, and hybrid. The latter refers to combinations of methods from the first two groups. In this section, we outline the main characteristics of two families of methods, namely the Kubelka-Munk and the Monte Carlo methods, which are representative of the deterministic and nondeterministic groups, respectively.

3.2.1 The Kubelka–Munk methods

Early in the twentieth century, Kubelka and Munk [150] developed a simple relationship between the absorption coefficient (μ_a) and the scattering coefficient (μ_s) of paint and its overall reflectance. This relationship is known as the Kubelka–Munk theory (henceforth referred to as K-M theory). Although it was originally developed for paint, it allows a simple quantitative treatment of the spectral properties of different materials.

The K-M theory applies energy transport equations to describe the radiation transfer in diffuse scattering media using two parameters: the scattering and the absorption coefficients. This theory also assumes that the medium presents inhomogeneities which are small compared with its thickness. As originally stated, it is considered to be a two-flux theory because only two types of diffuse radiant fluxes are involved, namely a diffuse downward flux, Φ_{d_j}, and a diffuse upward flux, Φ_{d_i}. The relations between the fluxes are expressed by two simultaneous linear differential equations [150]. Before presenting these equations, however, we should examine the passage of light through a material elementary layer whose thickness is denoted by dh (Figure 3.7).

Although dh is small compared with the thickness of the material, denoted by h, it is assumed to be larger than the diameter of particles (pigments or dyes)

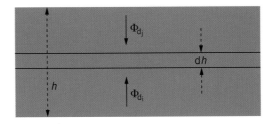

FIGURE 3.7

Geometry used to formulate the Kubelka–Munk two-flux (Φ_{d_j} and Φ_{d_i}) differential equations with respect to an elementary layer (thickness equal to dh) of a given material (thickness equal to h).

embedded in the material. Hence, only the average effect of the pigments on Φ_{d_j} is taken into account [135]. More specifically, the presence of pigment reduces the diffuse radiant fluxes. This reduction is because of absorption, given by $\mu_a \Phi_{d_j} dh$, and scattering (reversal of direction), given by $\mu_s \Phi_{d_j} dh$. Similarly, the upward flux Φ_{d_i} is also reduced because of absorption, given by $\mu_a \Phi_{d_i} dh$, and scattering, given by $\mu_s \Phi_{d_i} dh$. Furthermore, the amount $\mu_s \Phi_{d_j} dh$ is added to ϕ_{d_i}, and the amount $\mu_s \Phi_{d_i} dh$ is added to Φ_{d_j}.

The resulting differential equations for the downward and upward fluxes are given by

$$-d\Phi_{d_j} = -(\mu_a + \mu_s)\Phi_{d_j} dh + \mu_s \Phi_{d_i} dh \qquad (3.15)$$

and

$$d\Phi_{d_i} = -(\mu_a + \mu_s)\Phi_{d_i} dh + \mu_s \Phi_{d_j} dh, \qquad (3.16)$$

respectively. Equations 3.15 and 3.16 can be integrated to provide the reflectance and transmittance of the material [279].

The original K–M theory can be seen as a two-parameter generalization of the one-parameter Beer's law (Section 2.6.1), i.e., if the scattering coefficient is equal to zero, it degenerates to Beer's law. However, if the absorption coefficient is equal to zero, the K–M theory gives fluxes that are linearly dependent on the distance traveled by light within the material [97].

Several extensions to the original two-flux formulation, such as the incorporation of additional upward and downward collimated fluxes (Chapter 5), have improved the accuracy of the results obtained using K–M-based methods. We remark, however, that this approach represents an approximation for the general radiative transfer (light transport) problem, and relative faster results are usually obtained at the expense of simplified material descriptions.

3.2.2 Monte Carlo methods

Monte Carlo methods are usually applied in conjunction with ray optics techniques to simulate light transport processes associated with light and matter interactions. More specifically, these radiative transfer processes are stochastically simulated as random walks by keeping track of photon (ray) histories as they are scattered and absorbed within a given material. There are, however, different strategies that can be applied in such simulations. At one end of this list, there are Monte Carlo-based models that simulate light transport by geometrically representing the internal structures of a material (e.g., cells [102]). At the other end, there are models that use the intuitive concept of layers in which the properties of the internal structures of the material are globally described through mathematical expressions (e.g., using precomputed functions to represent the bulk scattering of light within a given layer [204]). Alternatively, it is also possible to compute the light propagation profile within a material without resorting to either of these approaches, i.e., the directional changes are computed on the fly taking into account the geometrical characteristics of the individual structures of the material without explicitly storing them [140].

In order to illustrate the main steps of a Monte Carlo simulation procedure, we will consider the propagation of light within a given multilayered material (Figure 3.8). In this abridged Monte Carlo presentation, we will use independent random numbers, denoted by ξ_i (where $i = 1, 2, \ldots$), uniformly

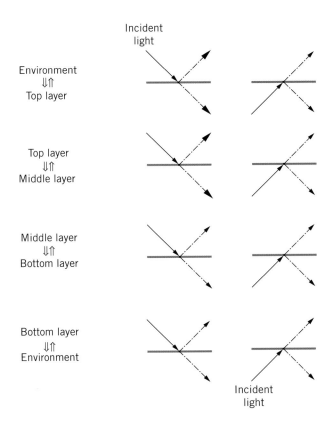

FIGURE 3.8

Sketch illustrating the main states (left) of a Monte Carlo simulation of light propagation within a multilayered material as well as the transitions affecting the downward (middle) and upward (right) ray trajectories.

distributed in the interval $[0, 1]$. Recall that a ray path can be represented by a random walk. In this case study, the internal dielectric interfaces of the material can be seen as states of this random walk whose transition probabilities are associated with the Fresnel coefficients (Section 2.2) computed at each interface. After the Fresnel coefficient F_R is computed at an interface, it is compared with a random number ξ_1 computed on the fly. If $\xi_1 \leq F_R$, then a reflected ray is generated; otherwise, a transmitted ray is generated. The random walk is terminated when the ray is either absorbed or propagated back to the environment.

A ray reflected or transmitted at an interface may enter a layer (Figure 3.9). In this case, its direction of propagation may be perturbed in a wavelength

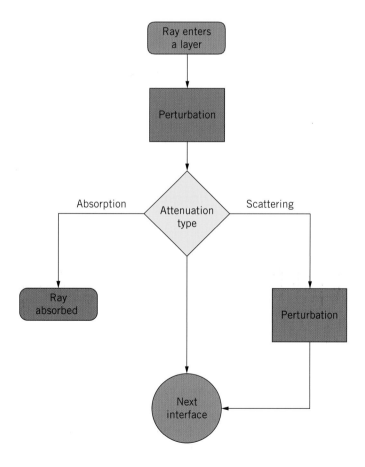

FIGURE 3.9

Flowchart illustrating the main steps of a Monte Carlo simulation of light propagation within a layer of a material.

independent fashion as a result of its collision with different material constituents (e.g., cells and organelles) whose dimensions are much larger than the wavelength of light. Accordingly, such a perturbation is performed using a warping function derived from a PDF that approximates the bulk scattering characteristics of the constituents of these materials (Section 3.1.2). For example, in the case of an irregularly shaped or rough elements that diffusively propagate light, the warping function provided by Equation 3.11 can be used.

In its passage through a layer, a ray may also be attenuated in a wavelength-dependent fashion. Considering a layer of thickness h and a polar propagation angle θ different from zero, the probability of such an attenuation is given by

$$\mathcal{P}_\mu(\lambda) = 1 - \epsilon^{-\mu(\lambda)h\sec(\theta)}. \tag{3.17}$$

Recall that the attenuation coefficient μ corresponds to the sum of the absorption coefficient (μ_a) and the scattering coefficient (μ_s) of the layer. The absorption coefficient can be obtained by multiplying the specific absorption coefficient (s.a.c.) of each layer constituent (e.g., pigments) by its concentration in the layer (Section 2.3.3). In the case of water, such a multiplication is not required. The scattering coefficient depends on the type of scattering phenomenon occurring in the layer. For example, in the case of a layer characterized by the presence of Rayleigh scatters (Section 2.3.2) with a refractive index η_s, the scattering coefficient can be computed using the following expression [168, 242]:

$$\mu_s(\lambda) = \frac{8\pi^3}{3\vartheta_s\lambda^4}(\eta_s(\lambda)^2 - 1)^2. \tag{3.18}$$

The probability $\mathcal{P}_\mu$ is then compared with a random number ξ_2. If $\mathcal{P}_\mu < \xi_2$, then neither absorption or scattering occurs. Otherwise, we need to determine the attenuation type. For this purpose, the absorption probability can be computed as follows:

$$\mathcal{P}_{\mu_a}(\lambda) = \frac{\mu_a(\lambda)}{\mu(\lambda)}. \tag{3.19}$$

Hence, if $\xi_3 \leq \mathcal{P}_{\mu_a}$, the ray is absorbed; otherwise, the ray is scattered.

Alternatively, the absorption testing can be seen as a stochastic interpretation of the Beer–Lambert law or one of its generalizations (Section 2.6.1). This interpretation indicates that the probability of absorption of a photon (ray) traveling a distance Δh at a certain wavelength λ in the medium is given by

$$\mathcal{P}_{\mu_a}(\lambda) = 1 - \epsilon^{-\mu_a(\lambda)\Delta h\sec(\theta)}. \tag{3.20}$$

By inverting Equation 3.20, the following expression for the estimation of the mean free path length p of a slant ray traveling in a pigmented tissue layer is obtained as

$$p(\lambda) = -\frac{1}{\mu_a(\lambda)} \ln(\xi_4) \cos(\theta). \qquad (3.21)$$

Similarly, if $p(\lambda) < h$, then the ray is absorbed; otherwise, it is scattered.

If the ray is scattered, the new direction is determined according to a function describing the specific type of scattering in effect within the layer. For example, in the case of Rayleigh scattering (Section 2.3.2), the new direction is determined according to the Rayleigh phase function given by Equation 2.24. Instead of applying the importance sampling technique presented earlier (Section 3.1.1), one can use a simpler, albeit less efficient, Monte Carlo technique called rejection sampling [108]. Using rejection sampling, one repeatedly generates the polar perturbation angle:

$$\alpha_R = 2\pi \xi_4, \qquad (3.22)$$

and accepts it only when

$$\xi_5 \leq 0.5(1 + \cos^2 \alpha_R). \qquad (3.23)$$

Because the directional perturbation in the azimuthal direction is symmetric (Section 2.5), the azimuthal perturbation angle is simply given by $\beta_r = 2\pi \xi_6$. As a result, the new ray direction is obtained by perturbing the ray according to the angular displacements given by α_r and β_R. Recall that phase functions (Section 2.5) represent photon and single particle interactions. Hence, the use of phase functions to specify the radiant intensity of thicker samples is only descriptive [126], i.e., they are used to provide a qualitative approximation for the bulk scattering of the materials within a Monte Carlo integration framework.

Monte Carlo models have been extensively used to simulate light interaction with inorganic [44, 140] and organic materials [204, 254] as they can provide a flexible, yet, rigorous approach to this problem [274]. These models can be easily implemented, and they are sufficiently flexible to allow the simulation of complex materials. Theoretically, Monte Carlo solutions can be obtained for any desired accuracy [201]. In practice, the accuracy of Monte Carlo simulations is bounded by the accuracy of the input parameters and the use of proper representations for the mechanisms of scattering and absorption of photons (rays). Furthermore, typically in Monte Carlo simulations, many trials (sample rays) are required to determine the overall local light transport behavior of a given material, which makes Monte Carlo models computationally expensive.

3.3 TECHNIQUES FOR MODEL EVALUATION

In order to assess the predictive capabilities of a local light transport model, one should compare its outputs with actual measured data. Although some models provide the reflectance and transmittance as the output of a given specimen, others provide its BDF. There are also a few comprehensive models that provide both sets of radiometric data. Actual measurements of reflectance and transmittance are performed using spectrophotometers, and actual measurements of BRDF and bidirectional transmittance distribution function (BTDF) are performed using goniophotometers [119, 135]. These devices are important basic tools for fundamental research in a myriad of fields, from remote sensing to biomedical optics. In this section, we discuss the computer simulations of such devices, henceforth called virtual measurement devices. Besides their application on the evaluation of local lighting models, these devices can also be used to gather data from previously validated models.

3.3.1 Actual and virtual spectrophotometry

A spectrophotometer is defined as an instrument for measuring the spectral distribution of reflected and transmitted radiant power, and spectrophotometry is defined as the quantitative measurement of reflection and transmission properties as a function of wavelength [77]. Figure 3.10 shows examples of typical spectrophotometric records.

Integrating spheres are used to provide reflectance readings where either the illuminant or viewing specification is "total" (hemispherical) or "diffuse only" [279]. In the latter case, a gloss trap may be incorporated in the design of an integrating sphere to reduce the influence of the specular component of specimens with mixed reflection behavior (Figure 3.11). Transmittance measurements are often carried out with sphere-type spectrophotometers as well [16]. In these measurement instances, however, the specimen is usually placed at the port of (light) entrance of the instrument (Figure 3.11).

The precision of a spectrophotometer is estimated by the ability of the instrument to replicate a measurement for a given specimen under the same

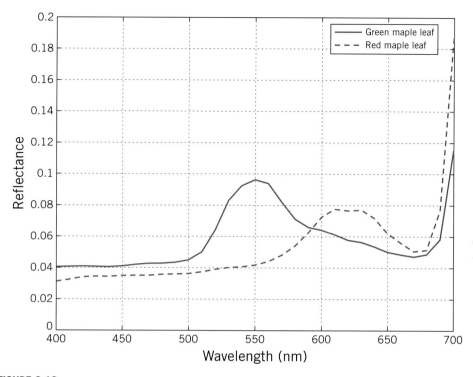

FIGURE 3.10

Measured (diffuse) reflectance spectra for a green maple leaf and a red maple leaf. The original data was obtained from the North Carolina State University (NCSU) spectra database [270].

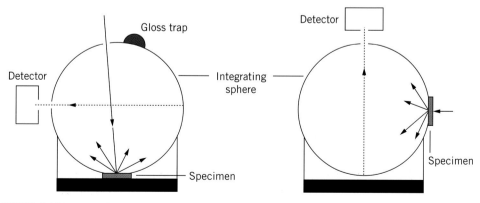

FIGURE 3.11

Diagrams illustrating the use of an integrating sphere to measure the reflectance (on the left) and the transmittance (on the right) of a given specimen.

spectral and geometrical conditions [135]. Well-designed and carefully calibrated spectrophotometers can yield results from the same specimen that differ from one measurement to the next. These differences, or uncertainties, are caused by variations in the components of the instrument, fluctuations in environmental conditions and changes in the specimen-handling procedure. In theory, a spectrophotometer is considered to be of high precision if the spectral measurements have an uncertainty, denoted by v, of approximately ± 0.001 [135, 163]. This means that at one time the device may read, for instance, a reflectance value equal to 0.572, but at other times it may read values as low as 0.571 or as high as 0.573. In practice, however, spectrophotometers usually have an absolute precision between 0.993 and 0.995, i.e., an uncertainty between ± 0.007 and ± 0.005 measurement units [288]. The accuracy of a spectrophotometer is measured by the ability of the device to provide, for a given set of illuminating and viewing geometries, the true spectral reflectance and transmittance of a given specimen, apart from random uncertainties occurring in repeated measurements [135].

Emitters and specimens used in actual measurements usually have circular areas, which can be represented by disks with radii r_1 and r_2 separated by a distance D (Figure 3.12). A spectrophotometer with an integrating sphere is simulated by sending (or shooting) sample rays from the emitter toward the specimen. These rays arrive at the specimen through a solid angle, ω_i, in the

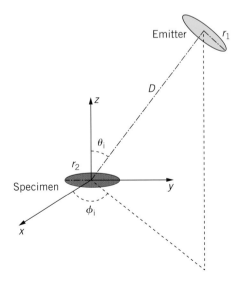

FIGURE 3.12

Sketch of a virtual spectrophotometer depicting an emitter (represented by a disk of radius r_1) and a specimen (represented by a disk of radius r_2) separated by a distance D.

direction of incidence ψ_i, which can be represented by a pair of spherical coordinates (ϕ_i, θ_i) (Figure 3.12).

Considering a ray optics stochastic simulation framework in which n rays are shot toward the specimen at a given wavelength λ, we can assume that each ray carries the same amount of radiant power, denoted by Φ_{ray}. If the total radiant power to be shot is Φ_i, then the radiant power carried by each ray is given by

$$\Phi_{ray}(\lambda) = \frac{\Phi_i(\lambda)}{n}. \tag{3.24}$$

Recall that reflectance describes the ratio of reflected power to incident radiant power, and transmittance describes the ratio of transmitted radiant power to incident power (Section 2.6.1). Considering this ratio, if n_r rays are reflected toward the upper hemisphere Ω_r, the directional-hemispherical reflectance of the specimen with respect to a given wavelength λ will be given by

$$\rho(\lambda, \omega_i, \Omega_r) = \frac{n_r}{n}. \tag{3.25}$$

Therefore, because one can simply count the number of rays reflected to the upper hemisphere to determine the reflectance of a specimen, a virtual spectrophotometer does not need to explicitly incorporate an integrating sphere to collect the reflected rays. The directional-hemispherical transmittance of a specimen is calculated in a similar manner, i.e., by counting the number of rays transmitted to the lower hemisphere.

This directional-hemispherical setup is usually used when a virtual spectrophotometer is used to obtain data from a previously validated model. In this case, one can use assume collimated rays, i.e., the sample rays have the same origin and hit the specimen at the same point. We remark, however, that for applications involving comparisons with actual measurements, the actual measurement conditions must be reproduced as faithfully as possible. In these situations, a conical-hemispherical geometry is usually used (Section 2.6.1).

A conical-hemispherical measurement geometry requires the generation of sample rays angularly distributed according to the spatial arrangement of the surfaces used to represent the emitter and the specimen. As mentioned by Crowther [57], the incident radiation from an emitter shows no preference for one angular region over the other. So, in order to simulate these measurement conditions, the origins and targets of the rays are random points (or sample points) chosen on the surfaces (e.g., disks) used to represent the emitter and the specimen, respectively. There are many sampling strategies that can be used to select the sample points on these surfaces [227]. For example, one can use a strategy based on the classical Monte Carlo stratified or jittered sampling [225]. This strategy uses a warping transformation to guarantee that the sample points are reasonably equidistributed on a disk of radius r_{disk}, and enables the computation of the pair (x,y) through the following warping function:

$$(x,y) = (r_d \cos(\Theta_d), r_d \sin(\Theta_d)), \tag{3.26}$$

where r_d and Θ_d correspond to radial and angular displacements, respectively. These displacements are stochastically computed (as a function of random numbers, ξ_1 and ξ_2, uniformly distributed in $[0,1]$) using:

$$(\Theta_d, r_d) = \left(2\pi\xi_1, r_{disk}\sqrt{\xi_2}\right). \tag{3.27}$$

After generating the x and y coordinates of a sample point, the z coordinate is added. For a sample point on the specimen, z is equal to zero, and for a sample point on the emitter, z is equal to the distance D between the disks (Figure 3.12). This distance, in turn, corresponds to the radius of the integrating sphere of a real spectrophotometer. Finally, to obtain the origin of a sample ray, the corresponding sample point (x,y,z) on the emitter shall be rotated according to a specified incidence geometry given by ϕ_i and θ_i (Figure 3.12).

According to the Bernoulli theorem [259], using a sufficiently large number of sample rays, one will have a high probability to obtain estimates within the region of asymptotic convergence of the expected value of reflectance, or transmittance, being measured. However, the processing time grows linearly with respect to the total number of sample rays because the cost of the algorithm is constant per ray. In order to minimize these computational costs, Baranoski et al. [25] proposed a bound on the number of sample rays required to obtain asymptotically convergent spectrophotometric values. This bound, which is derived from the *exponential* Chebyshev inequality [228], is given by

$$n_{sp} = \left\lceil \frac{\ln\left(\frac{2}{\varphi}\right)}{2\upsilon^2} \right\rceil, \tag{3.28}$$

where φ is the confidence on the estimation and υ is the uncertainty of the real spectrophotometer.

For example, considering a confidence of 0.01 and an uncertainty of 0.005, approximately $10^{5.02}$ rays are required to obtain reflectance and transmittance readings within the region of asymptotic convergence as illustrated by the reflectance estimates presented in Figure 3.13.

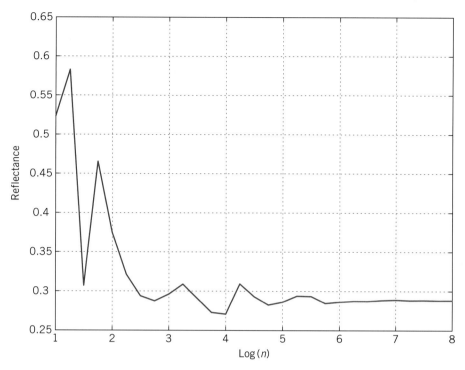

FIGURE 3.13

Directional-hemispherical reflectance estimates obtained for a dielectric material ($\eta = 2.5$) through a Monte Carlo simulation algorithm and considering different ray densities.

3.3.2 Actual and virtual goniophotometry

A goniophotometer is defined as an instrument that measures flux (power) as a function of angles of illumination and observation [77]. The light flux incident on the specimen comes from an emitter, and it is captured by a detector (photometer) after being reflected or transmitted by the specimen. For BRDF measurements, the detectors are placed on the hemisphere above the specimen; for BTDF measurements, the detectors are placed on hemisphere below the specimen. In the remainder of this section, we will examine the procedures related to BRDF measurements, implicitly addressing the procedures related to BTDF measurements by analogy.

It is important to note that goniophotometric measurements can be performed using different illumination and detection (viewing) geometries, and, as a result, there are many possible configurations for these devices. In order to obtain a complete goniophotometric record for a simple specimen, it would be necessary to perform a large number of measurements because the emitter and the detector would have to be moved independently of one another to every position on the hemisphere [135]. For many specimens, however, the most informative goniophotometric data are taken in the plane containing the direction of light incidence and the normal of the specimen. Many actual goniophotometers are abridged to this extent.

Similar to spectrophotometers, the accuracy of goniophotometers is also estimated by the ability of the instrument to replicate a measurement for a given specimen under the same spectral and geometrical conditions [135]. According to data provided in the measurement literature [94], the uncertainty of actual goniophotometers is usually around 0.5% or more.

In a virtual goniophotometer, radiance detectors can be represented by the patches of a collector sphere placed around the specimen. Figure 3.14 presents a sketch showing the principal components of a virtual goniophotometer and their geometrical arrangement. The light flux incident on the specimen comes from the emitter through a patch i, and the reflected light flux is collected by the detector covering a patch r. Both the illuminating and viewing directions can be varied independently within the hemisphere above the specimen. The position of emitter and patch i is given by the azimuth angle ϕ_i and the polar angle θ_i, respectively. The positions of the detector and patch r are given by the azimuth angle ϕ_r and the polar angle θ_r, respectively.

Using this arrangement, the BRDF for a direction associated with a given radiance detector placed in the upper hemisphere can be determined in terms of radiant power. More specifically, it is given by the ratio between the radiant

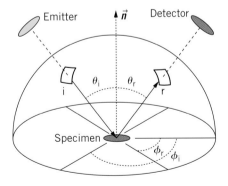

FIGURE 3.14

Sketch of a virtual goniophotometer. Patches **i** and **r** correspond to areas on the collector sphere associated with the illuminating (emitter position) and viewing (detector position) directions, respectively.

power reaching the detector, Φ_r, after interacting with the specimen, and the incident radiant power, Φ_i.

The corresponding expression used to compute the BRDF for light incident at wavelength λ, considering the solid angle in the direction of incidence, $\vec{\omega}_i$, and the solid angle in the direction associated with the radiance detector, $\vec{\omega}_r$, is given by

$$f_r(\lambda, \omega_i, \omega_r) = \frac{\Phi_r(\lambda)}{\Phi_i(\lambda)\omega_r{}^p}, \tag{3.29}$$

where the projected solid angle with respect to the direction associated with the radiance detector (patch r) is defined as

$$\omega_r{}^p = \frac{A_r \cos\theta_r}{D_r^2}, \tag{3.30}$$

where A_r is the area of patch r, D_r is the distance from the specimen to patch r, and θ_r is the angle between the direction associated with the radiance detector and the normal of the specimen.

Again, considering a ray optics stochastic simulation framework, BRDF estimates can be obtained by shooting n rays toward the specimen. The origins of the rays are random points uniformly chosen from a surface (e.g., a disk) used to represent the emitter, while the targets of the rays may also be represented by random points uniformly chosen from a surface used to represent the specimen. Assuming that each ray carries the same amount of power (Equation 3.24), the radiant power reaching the radiance detector (patch r) can be written as

$$\Phi_r(\lambda) = n_p \Phi_{ray}(\lambda), \tag{3.31}$$

where n_p is the number of rays hitting patch r.

Thus, replacing Equations 3.30 and 3.31 in Equation 3.29, the expression to compute the BRDF reduces to

$$f_r(\lambda, \omega_i, \omega_r) = \frac{n_p}{n \, \omega_r{}^p}. \tag{3.32}$$

Similarly, the BTDF is calculated considering radiance detectors placed on the lower hemisphere.

Krishnaswamy et al. [149] examined the implementation of virtual goniophotometric devices focusing on the subdivision of the collector sphere and on the ray density required to obtain asymptotically convergent BRDF and BTDF estimates. Their experiments indicated that the use of a subdivision

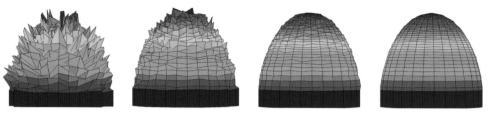

FIGURE 3.15

BRDF of a perfect diffuse material represented by three-dimensional plots obtained using different ray densities. From left to right: 10^5, 10^6, 10^7, and 10^8 rays.

technique based on equal project solid angles may, in general, provide a more uniform convergence for the estimates. Furthermore, similarly to the virtual spectrophotometer case, an upper bound for the number of rays required to obtain asymptotically convergent goniophotometric records can also be derived from the exponential Chebyshev inequality [228]. This bound depends on the number of patches, denoted by m, in which the collector hemisphere is subdivided. This bound is given by

$$n_{go} = m \left\lceil \frac{\ln\left(\frac{2}{\delta}\right)}{2v^2} \right\rceil. \tag{3.33}$$

For example, considering a confidence of 0.01, an uncertainty of 0.005, and a collector hemisphere subdivided into 900 patches, at most 10^8 rays would be required to obtain asymptotically convergent BRDF estimates as illustrated by the three-dimensional plots presented in Figure 3.15.

3.4 COLOR CONVERSION

The light that reaches our eyes corresponds to a spectral signal (color stimulus), which is reduced to three-dimensional color by our visual system. Ideally, spectral information should be preserved as long as possible in the rendering pipeline and only converted to three dimensions when mapped to a display device or a printing medium. In this section, we outline the basic steps involved in such a conversion.

Different color-specification systems can be used to map spectral information to color values [118]. They can be loosely divided into two groups. The first group is formed by device-independent systems such as the CIE (*Commission Internationale de L'Eclairage*) system, which allows the specification of color for any arbitrary color stimulus [188]. Because of its robustness and high precision, the CIE system is extensively used in industrial applications. The second group is formed by device-dependent systems such as the RGB color system. Besides being highly device specific, the RGB system lacks perceptual uniformity in the specification of colors. Despite these limitations, it is widely used in computer graphics applications because of its relative simplicity and straightforward implementation. For instance, the RGB color space is represented by a unit cube, and all possible RGB values correspond to attainable color, which facilitates color range checking [158].

In order to convert the spectral signal $S(\lambda)$ resulting from a rendering application to RGB values, one can use RGB tristimulus values, namely $\bar{r}(\lambda)$, $\bar{g}(\lambda)$, and $\bar{b}(\lambda)$. These can be obtained by converting CIE color-matching functions $\bar{x}$, $\bar{y}$, and $\bar{z}$ [48], using an appropriate transformation matrix T:

$$\begin{bmatrix} \bar{r}(\lambda) \\ \bar{g}(\lambda) \\ \bar{b}(\lambda) \end{bmatrix} = T \begin{bmatrix} \bar{x}(\lambda) \\ \bar{y}(\lambda) \\ \bar{z}(\lambda) \end{bmatrix}. \tag{3.34}$$

The matrix T is set according to the chromaticity and white point values associated with the target display (monitor) device [158]. For example, Table 3.1 provides the Society of Motion Picture and Television Engineers (SMPTE) values for phosphors' chromaticities and monitor white point.

The process used to obtain matrix T is straightforward, and it can be summarized as follows. Initially, a matrix A is formed

$$A = \begin{bmatrix} x_R & x_G & x_B \\ y_R & y_G & y_B \\ z_R & z_G & z_B \end{bmatrix}, \tag{3.35}$$

where each entry corresponds to a chromaticity coordinate (Table 3.1).

Table 3.1 SMPTE Chromaticity Coordinates and Wavelength Values

	x	y	z	**Wavelength (nm)**
Red	0.630	0.340	0.03	608
Green	0.310	0.595	0.095	551
Blue	0.155	0.070	0.775	455
White	0.313	0.329	0.358	

Using the coordinates of the monitor white point (Table 3.1), a vector $\boldsymbol{b}$ is then computed as

$$\boldsymbol{b} = A^{-1} \begin{bmatrix} \dfrac{x_W}{y_W} \\ 1 \\ \dfrac{z_W}{y_W} \end{bmatrix}, \tag{3.36}$$

and its elements will form the main diagonal of a matrix C:

$$C = \begin{bmatrix} \boldsymbol{b}_1 & 0 & 0 \\ 0 & \boldsymbol{b}_2 & 0 \\ 0 & 0 & \boldsymbol{b}_3 \end{bmatrix}. \tag{3.37}$$

The matrix T is finally given by

$$T = (AC)^{-1}. \tag{3.38}$$

After the computation of $\bar{r}(\lambda)$, $\bar{g}(\lambda)$, and $\bar{b}(\lambda)$, the tristimulus color (R, G, B) is quantified by sampling the spectral signal $S(\lambda)$ using the following equations:

$$R = \int_{400\,\mathrm{nm}}^{700\,\mathrm{nm}} S(\lambda)\bar{r}(\lambda)\mathrm{d}\lambda, \tag{3.39}$$

$$G = \int_{400\,\mathrm{nm}}^{700\,\mathrm{nm}} S(\lambda)\bar{g}(\lambda)\mathrm{d}\lambda, \tag{3.40}$$

and

$$B = \int_{400\,\mathrm{nm}}^{700\,\mathrm{nm}} S(\lambda)\bar{b}(\lambda)\mathrm{d}\lambda. \tag{3.41}$$

In practice, these integrations (Equations 3.39–3.41) are replaced by summations in which the spectral energy distribution represented by $S(\lambda)$ is discretized [107]. For example, considering only three wavelengths, these summations become

$$R = \sum_{i=1}^{3} S_i(\lambda)\bar{r}_i(\lambda), \qquad (3.42)$$

$$G = \sum_{i=1}^{3} S_i(\lambda)\bar{g}_i(\lambda), \qquad (3.43)$$

and

$$B = \sum_{i=1}^{3} S_i(\lambda)\bar{b}_i(\lambda). \qquad (3.44)$$

We remark that certain color-specification systems may be more suitable than others for a certain application, and different procedures can be used to convert values from one system to another. Hence, the above color-conversion outline should be seen as just one possible route that one can take to specify color. In fact, the selection of a color system appropriate for a given application is an important and complex task, and it has been the object of careful study in several fields, including computer graphics [29, 158, 241].

Bio-optical properties of human skin

4

Skin is an inhomogeneous organ with complex optical properties. In order to accurately simulate the mechanisms of light propagation and absorption that determine its visual attributes, it is necessary to carefully account for its biophysical and structural characteristics. In this chapter, we examine these characteristics and how they affect skin optics and ultimately its appearance, which may vary considerably not only among different individuals, but also among different cutaneous regions belonging to the same individual.

The appearance of human skin depends on spectral and spatial light distributions controlled by endogenous and exogenous factors. Endogenous factors are associated with skin constituents (e.g., pigments, cells, and fibers), whereas the exogeneous factors are associated with environmental conditions (e.g., illumination and temperature) and the presence of external materials (e.g., hair, oil, sweat, and cosmetics). The variations of these factors among the human population result in different spectral signatures and scattering profiles. Such variations and their outcomes are also reviewed in this chapter.

Several photobiological processes that affect the appearance and health of human skin are triggered by electromagnetic radiation outside the visible domain. Accordingly, we close this chapter with an overview of these phenomena and their visual and medical implications.

4.1 STRUCTURAL AND BIOPHYSICAL CHARACTERISTICS

An important issue to be considered in the simulation of light interactions with the human skin is the level of abstraction used to represent its different tissue layers. These layers have distinct characteristics not only in terms of their dimensions and cellular arrangements, but also in terms of their optical properties. Clearly, it may not be practical to consider all layers and sublayers resulting from either lack of data or computational complexity. Accordingly, the skin descriptions used in skin optics studies usually include three main layers, namely stratum corneum (SC), epidermis, and dermis, and sometimes a substrate layer known as hypodermis (Figure 4.1). In the remainder of this

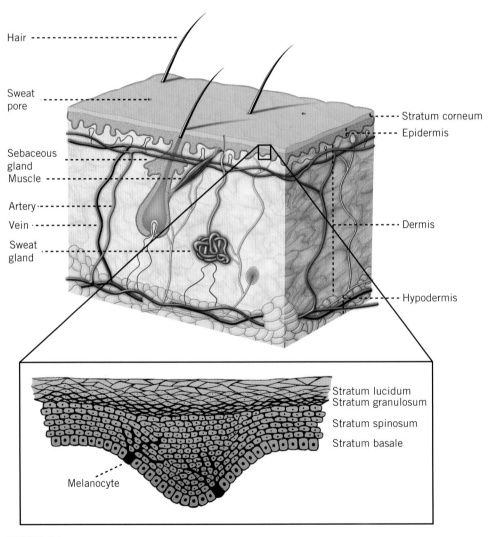

FIGURE 4.1

Diagram illustrating the stratified structure of human skin. The inset is a zoom in of the epidermis layer depicting its sublayers and the melanocyte cells.

section, we outline the main structural and biophysical characteristics of these layers and how they affect the propagation and absorption of light.

The first and outermost section of human skin is the SC, which is a stratified structure approximately 0.01–0.02 mm thick [3, 11, 170]. There are skin structural models, however, that consider it part of another tissue, namely the epidermis [254]. The SC is composed mainly of dead cells, called corneocytes, embedded in a particular lipid matrix [244]. Light absorption is low in this tissue, with the amount of transmitted light being relatively uniform in the visible region of the light spectrum [82].

The epidermis is a 0.027–0.15-mm thick structure [11, 66, 170] composed of four main layers (stratum basale, stratum spinosum, stratum granulosum, and stratum lucidum). The epidermis propagates and absorbs light. The absorption property comes mostly from a natural pigment (or chromophore), melanin. There are two types of melanin found in human skin: the red/yellow pheomelanins and a brown/black eumelanin [42, 248]. Most individuals synthesize a mixture of both types [4].

The biosynthesis of melanin, known as melanogenesis, occurs through a complex biochemical process [235], which is still not completely understood. It starts with the oxidation of an amino acid known as tyrosine [8]. A rate-limiting enzyme, known as tyrosinase, catalyzes the first two oxygenic reactions of the biochemical pathways that lead to the formation of one of the melanin pigments, eumelanin or pheomelanin [133]. Melanogenesis takes place in membranous organelles, called melanosomes, found in the long filaments of melanocyte cells. In healthy skin, these cells are located in the stratum basale (Figure 4.1). In this epidermis, we also find the keratinocytes that produce keratin. This substance, which absorbs light (predominantly in the ultraviolet domain [31, 145]), is found not only in the epidermal cells, but also in hair and nails. After a certain period (on the order of hours) of exposure to light, the melanocytes transfer the melanosomes to the keratinocytes, where the melanin lays like an umbrella over the nucleus and protects it from the incident radiation [144]. The epidermal melanin pigmentation (EMP) can be classified into two types: constitutive (baseline pigmentation) and facultative (induced) pigmentation.

The absorption spectra of eumelanin and pheomelanin are broad (Figure 4.2), with higher values for shorter wavelengths. The skin color is mostly associated with eumelanin [248]. The ratio between the concentration of pheomelanin and eumelanin present in human skin varies from individual to individual, with significant overlap between skin types [248]. Different studies report values between 0.049 and 0.36 [192]. The melanin-absorption level depends on how many melanosomes per unit volume are in the epidermis. Typically, the volume fraction of the epidermis occupied by melanosomes

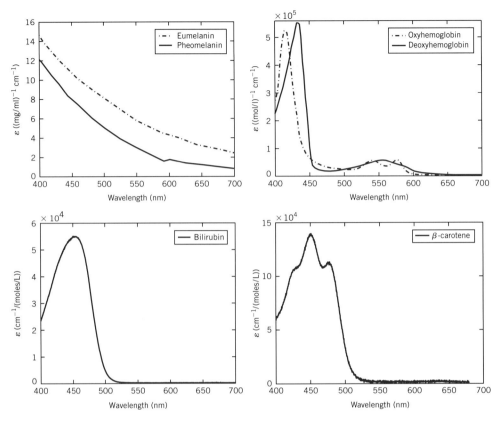

FIGURE 4.2

Spectral molar extinction coefficient curves for natural pigments (chromophores) present in the skin tissues. Top left: melanins [127]. Top right: hemoglobins [202]. Bottom left: bilirubin [203]. Bottom right: β-carotene [203].

varies from 1.3 (lightly pigmented specimens) to 43% (darkly pigmented specimens) [126]. The number of melanocytes present in various cutaneous regions of the human body may also vary significantly [235].

The dermis is a 0.6–3-mm thick structure [11, 66, 170], which also propagates and absorbs light. It can be divided into two regions: the papillary dermis and the reticular dermis. These layers are primarily composed of dense, irregular connective tissue with nerves and blood vessels (smaller ones in the papillary dermis and larger ones in the reticular dermis). The volume fraction of blood in human tissue can vary, roughly in the 0.2–7% range [91, 126]. The fluence rate of blood decreases as we get deeper into the skin, following an almost linear pattern in the dermis [264]. In the blood cells, we find another natural chromophore, hemoglobin, which absorbs light and gives blood its reddish color. Normally, the hemoglobin concentration in whole blood is between 134 and 173 g/L [281]. In the arteries, 90–95% of hemoglobin is oxygenated, and in the veins, more than 47% of the hemoglobin is oxygenated [12]. These two types of hemoglobin, namely oxygenated and deoxygenated hemoglobin, have slightly different absorption spectra (Figure 4.2). It worth noting that a mechanical, chemical, electrical, thermal, or luminous stimulation of the skin tissues can induce a reddening around the stimulus site due to a dilation of the blood vessels, which increases the volume fraction of blood in the dermis [1, 185]. This condition is known as erythema (Figure 4.3).

Other blood-borne pigments found in the dermis are bilirubin and carotenoids, notably β-carotene (Figure 4.2). These pigments contribute to the yellowish or olive tint of human skin. Bilirubin is a pigment derived from the degradation of hemoglobin during the normal and abnormal destruction of red blood cells. Bilirubin is normally filtered out of the blood by the liver. If this organ is not working properly, this process may be affected. In this

FIGURE 4.3

Photograph illustrating an erythema (redness) condition caused by a dilation of the blood vessels triggered by an environmental factor, in this case, overexposure to sunlight.

case, the excessive quantity of this pigment in the blood may cause a yellowness in the skin and eyes. This change in skin color is a visual symptom of a medical condition known as hyperbilirubinemia or jaundice [211], which is often observed in newborn babies (Figure 4.4). In newborn babies, however, the situation is slightly different. They normally have an elevated level of red blood cells. As these cells eventually breakdown, there is a spike in the bilirubin levels, and it takes some time for the liver to process it all out, resulting in jaundice. Because newborn babies do not have their blood-barrier in the brain fully formed yet, if these elevated bilirubin levels persist, this substance can eventually enter the brain and cause neuronal damage [214].

The pigment β-carotene is found in many species of plants, and their ingestion in relative large quantities may increase the amount of this pigment not only in the dermis, but also in the epidermis and SC [9, 152]. Other carotenoids, such as α-carotene, lutein, zeaxanthin, and lycopene, can also be found in human skin [239]. Among these carotenoids, lycopone occurs in

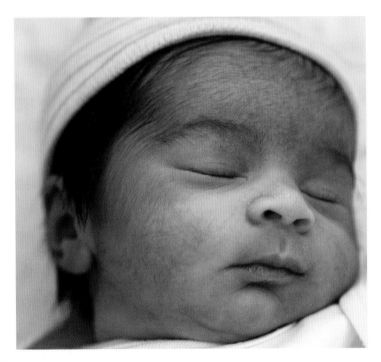

FIGURE 4.4

Photograph illustrating the characteristic jaundice (yellowness) condition caused by an excessive production of bilirubin, usually observed in newborn babies.

larger quantities. Its absorption spectra is similar to the β-carotene spectra, with a slight shifted toward the red end of the light spectrum [59]. It is important to note that the levels of carotenoids vary in the different areas of the skin, with larger levels occurring in the forehead, palm of the hand, and dorsal skin [239].

Other organic and inorganic materials, such as proteins, lipids, and water, also contribute to the absorption of light in human skin. Their net absorption effect provides the baseline absorption coefficient for human pigmentless skin tissues. Because of the low magnitude of this coefficient in the visible domain, compared with the absorption coefficients of the other absorbers present in human skin, skin optics researchers usually assumed that its contribution is negligible for applications aimed at the visible portion of the light spectrum [12, 214]. We remark, however, that absorption of light by these materials may be significant outside this spectral domain [125, 197], especially for tissue optics applications demanding a high level of accuracy.

The hypodermis (or subcutis) is formed by two types of tissues: the loose connective tissue and the subcutaneous adipose tissue [6]. Although the former is composed of blood, lymphatic vessels, and nerves, the latter consists mostly of white fat cells that are grouped together forming clusters. These cells contain smooth droplets of lipids called adipocytes [28]. The thickness of the hypodermis varies considerably throughout the body and among males and females from different age groups [6]. It can be up to 3 cm thick in the abdomen and absent in the eye lids. The absorption of visible light in the hypodermis is determined by oxyhemoglobin content, whereas the absorption of infrared light is determined by water and lipid content [28, 176].

4.2 SPECTRAL SIGNATURES

The process of light absorption by pigments present in the skin issues is responsible for the most noticeable skin spectral variations across the human population. Among these pigments, melanin plays a key role. Its biosynthesis is tied to the concentration of melanosomes in the epidermis, which is determined by genetic factors that also control other stages of the melanogenesis process [193]. Consequently, individuals from different ethnical groups will likely have different concentrations of melanosomes and, under neutral illumination conditions, will have different spectral signatures [7]. Figure 4.5 provides examples of different spectral signatures measured from skin specimens from various ethnical groups.

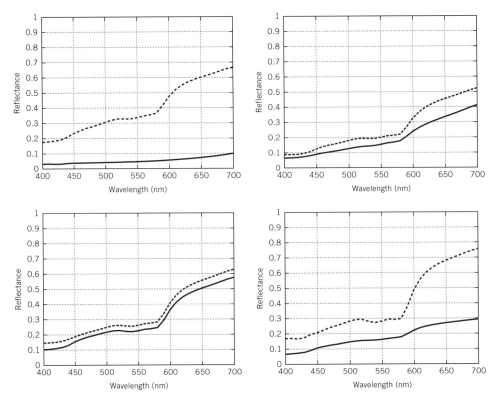

FIGURE 4.5

Measured (diffuse) reflectance spectra for skin specimens from individuals belonging to different ethnical groups. Top left: African Americans. Top right: East Indians. Bottom left: Asians. Bottom right: Caucasians. (The original data was obtained from the North Carolina State University (NCSU) spectra database [270]).

The skin of individuals with a genetic predisposition to a higher concentration of melanosomes usually tend to have a spectral signature characterized by lower reflectance values, which monotonically increase across the visible portion of the light spectrum. However, the skin of individuals with a genetic predisposition to a lower concentration of melanosomes tend to have a spectral signature characterized by higher reflectance values which vary considerably in the region between 500 and 600nm. The lower amount of melanin in this case makes the effects of light absorption by oxyhemoglobin to become more prominent. Hence, the variations in the absorption spectrum of oxyhemoglobin in this region (Figure 4.2) result in the characteristic "W" shape observed in the spectral signature of moderately or lightly pigmented skin specimens (Figure 4.5).

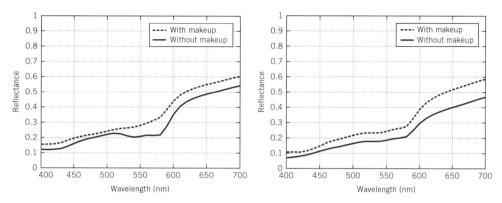

FIGURE 4.6

Measured (diffuse) reflectance spectra for Caucasian skin specimens with and without makeup. Left: lightly pigmented specimens. Right: moderately pigmented specimens. (The original data was obtained from the North Carolina State University (NCSU) spectra database [270]).

Different amounts of melanin not only affect the overall magnitude of the reflectance curves of individual from different ethnical groups, but also the magnitude of the reflectance curves of individuals from the same ethnical group. Besides endogenetic factors, such as the volume of melanosomes, exogenic factors, such as the presence of external materials (Figure 4.6) and the incidence of ultraviolet light, may also contribute to the relative wide range of spectral signatures that characterize the human population. These exogenic factors are further examined in the following sections.

4.3 SCATTERING PROFILES

The scattering profile of human skin has two main components: surface and subsurface scattering. Surface scattering is characterized by a dependence on refractive index differences and on the angle of incidence of the incoming light (Figure 4.7). It follows the Fresnel equations [65], and it is affected by the presence of folds in the SC (Figure 4.8). The aspect ratio of these mesostructures depends on biological factors such as aging and hydration [244, 247, 287], as well as the presence of external substances such as cosmetics and lotions (Figure 4.9). Approximately 5–7% of the light incident (over the entire spectrum) on the SC is reflected back to the environment [254].

The portion of the light that is not reflected on the skin surface is transmitted to the internal tissues. Besides the reflective–refractive scattering caused

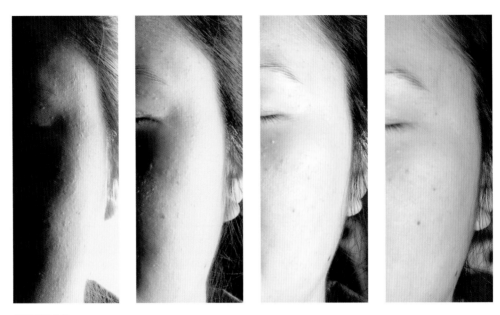

FIGURE 4.7

A sequence of photographs depicting light scattering on a skin surface as the illumination geometry changes. Left: grazing angle of incidence. Right: nearly normal angle of incidence.

FIGURE 4.8

Photograph depicting a close up of a skin surface.

FIGURE 4.9

Photograph illustrating the effects of external substances on the skin surface reflectance. Left: before lotion application. Right: after lotion application.

Table 4.1 Refractive Indices (Average Values) for the Stratum Corneum, Epidermis, Papillary Dermis and Reticular Dermis

Layer	Value	Source
Stratum corneum	1.55	[64]
Epidermis	1.4	[254]
Papillary dermis	1.36	[128]
Reticular dermis	1.38	[128]

by the refractive index differences at interfaces between different cellular layers (Table 4.1), two other types of scattering occur within the skin layers: Mie and Rayleigh scattering [126].

The SC and the epidermis are characterized as forward scattering media [38]. In the former, this behavior is because of the alignment of the fibers, whereas in the later, it is because of Mie scattering caused by particles (e.g., cell organelles) that are approximately the same size of the wavelength of light. Furthermore, the level of forward scattering for these tissues is wavelength dependent as demonstrated by the goniometric measurements performed by Bruls and van der Leun [38] for both the SC and the epidermis (Figure 4.10). Considering, for example, the fraction of energy transmitted within an angle of 22.5° of the normal of the sample at 546nm, it corresponded approximately to 83 and 59% for SC and epidermis samples, respectively. When these values are compared with the value computed for a diffusively scattering sample (14.6% within 22.5° from the normal), they clearly illustrate the forwardly scattering behavior of these tissues. A similar scattering behavior is observed with respect to ultraviolet light (Figure 4.10). The regions where melanosomes are being deposited in the epidermis are characterized by multiple scattering events. Melanosomes predominantly

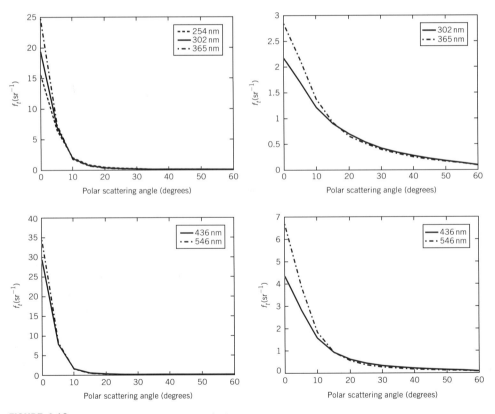

FIGURE 4.10

Graphs showing BTDF curves for stratum corneum (left) and epidermis (right) samples, considering ultraviolet (top) and visible (bottom) incident light. The curves were computed by Baranoski et al. [19, 21] using scattering data measured by Bruls and van der Leun [38].

exhibit forward scattering with respect to ultraviolet light, whereas degraded melanosomes (also called melanin dust) exhibit a more symmetric scattering profile in this spectral domain [42].

In the dermis, collagen fibers (approximately 2.8 μm in diameter and cylindrical [126]) are responsible for Mie scattering, whereas smaller scale collagen fibers located in the papillary dermis and other microstructures are responsible for Rayleigh scattering [126]. Light gets scattered multiple times inside the dermis before it is either absorbed or propagated to another layer. This means that the spatial distribution of the light scattered within the dermis quickly becomes diffuse [128], which contributes to the translucency of human skin as shown in Figure 4.11. Although Mie scattering produces variations on both ends of the visible region of the light spectrum, Rayleigh scattering, being inversely proportional to the wavelength of light, produces larger variations for shorter wavelengths (Section 2.3.2). The net effect of these two types of scattering follows an inverse relationship with respect to the wavelength of light travelling in the dermis. As a result, as observed by Anderson and Parish [11], longer wavelengths, notably in the infrared domain, tend to penetrate the dermis to a greater extent than shorter wavelengths. The amount of light that reaches the hypodermis is further scattered by the adipocyte and collagen fibers that form the adipose and connective tissues, respectively [28].

The net effect of skin surface and subsurface scattering is a reflective bidirectional reflectance distribution function (BRDF) intermediate between that

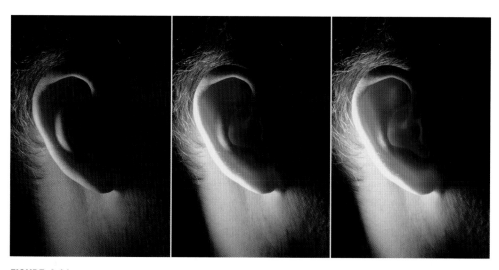

FIGURE 4.11

Photographs illustrating the diffuse pattern of transilluminated light emerging from skin tissues as the light intensity is increased.

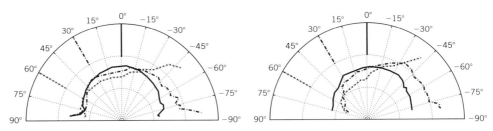

FIGURE 4.12

Measured skin BRDF profiles of Caucasian individuals. Left: adult male. Right: young female. Curves were obtained for three angles of incidence (0°, 30°, and 60°) and plotted considering the plane containing the direction of light incidence and the normal of the specimen [164–166].

expected from an ideal Lambertian reflector and an ideal specular reflector. This BRDF presents an angular dependence, and it becomes more diffuse for small angles of incidence as illustrated by the measurements provided by Marschner et al. [164–166] (Figure 4.12).

The appearance of healthy human skin may also be affected by the presence of congenital anomalies of the skin, such as moles and various types of birthmarks, termed nevi. Sometimes, as a result of papillary dermis becoming thin, light propagating in this tissue is absorbed by structures beyond the reticular dermis instead of being backscattered [49]. This effect results in areas characterized by a dark appearance due to excessive pigmentation. Other regions with a similar appearance may result from deposits of blood because of large concentrations of blood vessels.

4.4 INTERACTIONS WITH INVISIBLE LIGHT

Although our perception of skin appearance attributes takes place in the visible domain, several photobiological processes that affect these attributes are initiated by electromagnetic radiation in the ultraviolet and infrared domains. In this section, we examine these processes and discuss relevant implications not only for the simulation of skin appearance but also for the visual and noninvasive diagnosis of skin diseases.

4.4.1 Ultraviolet domain

According to the CIE, ultraviolet (UV) radiation can be divided into three regions [27]: UV-A (ranging from 315 to 380 nm), UV-B (ranging from 280 to 315 nm), and UV-C (ranging from 100 to 280 nm). Ultraviolet light can induce a myriad of photobiological processes such as erythema, melanogenesis photoaging (discoloration and wrinkle formation), phototoxicity, photoirritation, and photoallergy [30, 139, 144, 193, 263]. The facultative EMP that is produced following exposure to UV radiation varies depending on the spectral distribution of the light source [144]. UV-C is mostly absorbed by the ozone layers in the atmosphere. UV-B penetrates deeper than UV-C in skin layers, and it may increase melanogenesis after a certain period (6–8 h) that follows an erythema reaction [144]. It is responsible for the development of tan (Figure 4.13) in individuals with certain skin phototypes (SPTs). These phototypes were established to classify the variable sensitivity of different people exposed to the same dose of UV-B at identical sites. Their different responses to sunlight exposure are determined by genetic factors that control the formation and transfer of melanosomes from melanocytes to keratinocytes. This classification, originally proposed by Fitzpatrick [89], is based on the amount of constitutive color of skin (genetically produced epidermal melanin content in regions of the body that are not normally exposed to sunlight), and on the capacity of skin to darken or tan as a result of sun exposure [193]. The SPTs are quantified in terms of the minimum erythema dose (MED). It represents the amount of radiant energy necessary to produce the first perceptible

FIGURE 4.13

Photograph illustrating the effects of tanning in an area close to a skin region not normally exposed to sunlight.

and unambiguous development of redness, with clearly defined outlines, interpreted from 16 to 24 h after exposure to ultraviolet radiation [208]. The MED may vary from 20 to 25 mJ/cm^2 for a SPT I individual to 80–120 mJ/cm^2 for a SPT VI individual.

According to Pathak [193], the SPTs and their reactivity to sunlight can be summarized as follows:

- SPT I
  - constitutive skin color: white
  - responses: always burns easily, no immediate pigment darkening reaction (IPD), never tans, and usually peels

- SPT II
  - constitutive skin color: white
  - responses: always burns easily, trace IPD, darkening of freckles, tans minimally, and usually peels

- SPT III
  - constitutive skin color: white
  - responses: burns moderately, some IPD, tans gradually and uniformly (light brown)

- SPT IV
  - constitutive skin color: light brown
  - responses: burns lightly, significant IPD, always tans well (moderate brown)

- SPT V
  - constitutive skin color: brown
  - responses: burns minimally, substantial IPD, tans profusely (dark brown)

- SPT VI
  - constitutive skin color: brown
  - responses: rarely burns, substantial IPD, tans profusely and deeply (black)

UV-A penetrates deeper than UV-B [154], and it can induce epidermal pigmentation immediately with exposure [144]. There are only a few exogenous chemical substances that can absorb UV-A efficiently [235]. Commercial tanning beds emit mostly ultraviolet light in the UV-A range. Although the lack of UV-B in infants and small children may lead to disruption of bone growth and increase the probability of tooth decay [235], overexposure to ultraviolet radiation can induce the formation of skin lesions, such as carcinomas

[193] and melanomas [90]. The latter is the most serious form of skin cancer as it presents a high potential for metastasis (transference of cancerous cells to other parts of the body) and low cure rates [106]. Not surprisingly, a substantial amount of research and resources are applied in the development of effective sunscreens [235].

More than two billion people in the world have SPTs V and VI, which are associated to a relative low incidence of skin cancer. Individuals with SPT III and IV have the full ability for tanning and they are called melanocompetent. Individuals with SPT I and II have a limited or no capacity to tan after exposure to UV radiation, and they are referred to as melanocompromised. This "inability to tan" is considered one the most important risk factors for the development of cutaneous melanoma [90].

The appearance of pigmented skin lesions is usually described in terms of two parameters: mean reflectance and variegation [258]. The former corresponds to the capacity of the lesion to reflect light. The latter indicates how evenly distributed is the light reflected by the lesion, and it is associated to the distribution of pigments, notably melanin and hemoglobin. Melanomas usually present a lower reflectance and are more variegated than other lesions, especially in the red domains and near-infrared domains [258].

Melanoma occurs when melanocytes reproduce at an abnormal rate. If they remain in the epidermis, there are no significant differences with respect to other skin lesions that appear as patches of dark color (usually because of excessive concentration of melanin). This type of melanoma, known as in-situ melanoma, is not life threatening. If the malignant melanocytes penetrate the dermis, however, they leave deposits of melanin. This process results in spectral signatures with characteristics markedly different from the characteristics normally observed in healthy skin or in benign lesions. The probability of metastasis increases with the depth of melanocyte penetration, and the disease prognosis becomes increasingly worse [49].

The classification of skin lesions through reflectance measurements is still an open problem because of the inherent complexities of the optical processes that affect their appearance attributes. For example, the excessive pigmentation characteristic of certain benign lesions (e.g., common nevi) can also result from a reduction of papillary dermis thickness [49]. Also, the dermal melanin found in melanomas can, sometimes, be found in benign lesions (e.g., junctional nevus and blue nevus) [49].

4.4.2 Infrared domain

The primary source of infrared radiation is heat or thermal radiation (Section 2.3.1), i.e., the radiation produced by the motion of atoms and

Table 4.2 Water Content (Average Values) for the Stratum Corneum, Epidermis, Papillary Dermis and Reticular Dermis [171]

Layer	Value (%)
Stratum corneum	5
Epidermis	20
Papillary dermis	50
Reticular dermis	70

molecules in an object. The higher the temperature, the more the atoms and molecules move and the more infrared radiation they produce. Humans, at normal body temperature (around 35°C or 308 K), radiate most strongly in the infrared domain. In fact, for infrared radiation, the human body is a very good approximation of an ideal blackbody radiator irrespective of skin pigmentation [219]. This biophysical characteristic of the human body has motivated the application of infrared thermography in medical diagnosis [5]. Infrared thermography uses infrared imaging and measurement devices to produce visual representations of thermal energy emitted from a material. Medical applications of infrared thermography include the use of skin temperature as an indicator of subcutaneous pathological processes [184].

Other applications in the infrared domain include the noninvasive diagnosis of skin lesions [177], measurement of blood glucose [252], and quantification of hemoglobin oxygenation [171], as well as the remote sensing of human skin, which involves the deployment of hyperspectral cameras to aid in search and rescue missions [187]. One key aspect to be considered in simulations of light and skin interactions in this domain is the absorber role of water. Although in the visible domain, the specific extinction coefficient of water is almost negligible in comparison with other absorbers present in the skin tissues, in the infrared domain, it is quite pronounced (Figure 2.3). Hence, the presence of water in the skin tissues needs to be accounted for in these simulations. Table 4.2 provides average values for water content in different skin layers. These values, especially for SC, may vary considerably under different environmental conditions [2].

The incidence of infrared radiation can trigger a number of photobiophysical phenomena, which may affect the appearance of human skin. For example, it can increase the temperature of the tissues, and excessive exposure can induce erythema reactions.

4.4.3 Terahertz domain

Between the long-wavelength edge $(10^{-6}\,m)$ of far-infrared radiation and the low-wavelength edge $(10^{-3}\,m)$ of microwave radiation, one finds the Terahertz (THz) domain. The interactions of THz radiation with human skin are being studied with the purpose of developing algorithms for the detection of cancer and other tissue disorders [196, 278]. Similarly, the electromagnetic responses of skin to sub-THz radiation (in the millimeter and submillimeter range) are being studied for their possible use in the remote sensing of human photobiological responses [85]. Like in the infrared domain, water also has a key role in the light and skin interactions in the THz domain.

Simulations in health and life sciences

The simulation of light interaction with human skin has been an object of extensive research in many fields for decades. Long before the first local lighting models appeared in the computer graphics literature, detailed biologically based models of light transport in organic tissue were already available in the scientific literature. In fact, most of the models proposed by the computer graphics community were built on modeling techniques developed and employed in those fields. Hence, to fully assess the efforts made by the computer graphics community toward the predictive modeling of skin appearance, it is essential to revisit seminal works that provide the basis for this line of computer graphics research. Furthermore, the modeling of skin optical properties continues to be an active area of theoretical and applied research in life sciences. However, the contributions of computer graphics to these efforts are still marginal. One of the key steps to change this situation is to strengthen the correctness and fidelity of the computer graphics simulations, which can be achieved by addressing data constraints and open problems from an interdisciplinary perspective.

In this chapter, we provide an overview of relevant simulation approaches and models of light interaction with human skin aimed at applications in health and life sciences. We start by briefly addressing the context in which these models are inserted, i.e., the application requirements that have influenced their design. The body of work on the modeling of tissue optics is quite extensive, and a comprehensive survey on this topic would require a entire book devoted to it. Hence, our discussion focuses on models that have been primarily aimed at the simulation of light and skin interactions. These models are grouped and examined according to the methodologies employed in their design. For more general information on topics related to the modeling of skin

DOI: 10.1016/B978-0-12-375093-8.00005-8

optics, the reader is referred to the texts by Cheong et al. [45], Tuchin [254], and Störring [241].

5.1 SCOPE OF APPLICATIONS

The models of light interaction with human skin developed by the scientific community are usually designed to support the noninvasive measurement of tissue optical properties to be used in the diagnosis [245, 280], prevention [199], and treatment of skin diseases [62, 237]. Accordingly, a substantial portion of the modeling work done in these fields is either laser-based or aimed at wavelengths outside the visible region of the light spectrum (ultraviolet and infrared domains).

Models specifically developed for biomedical applications usually provide as output radiometric quantities, such as reflectance and transmittance, describing the spectral power distribution of skin tissues. None of these models was designed to output radiometric quantities, such as BRDF and BTDF, describing the spatial (directional) power distribution of these tissues. Some of these applications involve model inversion procedures [289]. An inversion procedure is a way to derive biochemical and optical properties from in situ and noninvasive measurements. These measurements, which often correspond to spectral reflectance and transmittance, are usually obtained by placing a sensor against or at some distance from the tissue. The term *inversion* in this context implies a reversal of the actual process of calculating reflection and transmission; i.e., using reflectance and transmittance values as input to an inverted model, one attempts to determine absorption and scattering properties of the tissues (Figure 5.1). Similar approaches are used in colorimetry to determine the relationship between color appearance and the content and distribution of various pigments [253]. Ideally, light transport models should be employed in inversion procedures after their correctness has been quantitatively verified through comparisons with measured data. In practice, to the best of our knowledge, several models used in such applications have been only qualitatively evaluated.

FIGURE 5.1

Sketch illustrating the general idea behind inversion procedures, i.e., a reflectance (ρ) and transmittance (τ) model is inverted and used to determine tissue optical parameters such as attenuation coefficient (μ) and albedo (γ).

Some medical applications, such as phototherapy for the treatment of skin inflammatory diseases or photodynamic therapy for the treatment of skin cancer, require in-vivo light dosimetry. Dosimetry corresponds to the measurement of radiant energy fluence rate, given in terms of irradiance (radiance integrated over all directions), and fluence (the time integral of the fluence rate) [237]. The in-vivo light dosimetry is usually performed either by inserting a probe in the tissue or by applying an inversion procedure.

5.2 KUBELKA–MUNK THEORY–BASED MODELS

The Kubelka–Munk (K–M) theory–based models, also known as K–M models or flux models [45], employed in biological tissue optics, use K–M equations relating tissue optical properties to measure reflectance and transmittance (Section 3.2.1). Although these models allow a rapid determination of optical properties through inversion procedures, their relative simplicity and speed are tied to their relative low accuracy. This aspect has been improved by adding more coefficients and fluxes to the original two-flux K–M formulation. For example, van Gemert and Star [265] included optical depth and effective albedo and phase function in their K–M model. The latter was used to approximate the tissue scattering behavior observed experimentally, and it consisted of a combination of two terms: one representing forward peaked scattering and another representing symmetric scattering. Tuchin et al. [256, 282] used a four-flux model composed of the two diffuse fluxes used in the original K–M theory and two fluxes represented by collimated laser beams, one incident and another reflected from the bottom boundary of the specimen (Figure 5.2). Yoon et al. [284, 286] used a seven fluxes model (Figure 5.3) to obtain a three-dimensional representation of the scattered radiation caused by an incident laser beam in a semi-infinite medium (infinite in x and y, but finite in z). They also incorporated a phase function composed of a symmetric term and a Henyey–Greenstein phase function (HGPF) term.

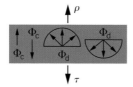

FIGURE 5.2

Sketch illustrating the four fluxes used in the model developed by Tuchin et al. [256, 282], namely two diffuse fluxes (Φ_d) used in the original K–M theory, and two fluxes (Φ_c) represented by collimated laser beams.

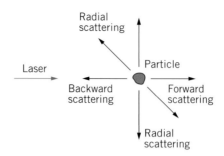

FIGURE 5.3

Sketch illustrating the seven fluxes considered in the model developed by Yoon et al. [284, 286].

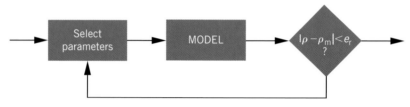

FIGURE 5.4

Flowchart illustrating the iterative process used by Doi and Tominaga [66] to obtain skin parameters by comparing the difference between the modeled reflectance (ρ) and the measured reflectance (ρ_m) with an error threshold (e_r).

In skin optics, the K–M theory was initially applied to specific skin tissues. Anderson and Parrish [11] used a K–M model to compute absorption and scattering coefficients for the dermis tissues. Wan et al. [271] extended this model to compute the absorption and scattering coefficients for the epidermis tissues and took into account both collimated and diffuse incident irradiance. In both cases [11, 271], the forward scattering in the epidermis was not considered. Diffey [64] proposed a K–M model which added two features to the previous models; namely, it takes into account forward and backward scattering and allows changes in the refractive index at the air and skin interfaces. Doi and Tominaga [66] presented a model which considers the skin is composed of two layers: epidermis and dermis. They apply the K–M theory to both layers. Their model provides weights for five skin pigments (melanin, β-carotene, oxyhemoglobin, deoxyhemoglobin, and bilirubin) as well as the skin surface reflectance. These six parameters are obtained by fitting the estimated reflectance to measured values (Figure 5.4) using the least squares method [39].

Cotton and Claridge [56] proposed a model to determine the color of human skin which applies a two-flux K–M formulation to the epidermis, papillary dermis, and reticular dermis. This model takes into account the presence of collagen, melanin, and blood. It was later incorporated into a spectrophotometric intracutaneous analysis (SIAscopy) framework [50], which is commercially known as SIAscope. This framework is mainly designed to derive information about the internal structure and composition (melanin, blood, and collagen contents) of pigmented [177, 245] and other skin lesions [167] from color images. Initially, a hand-held device (Figure 5.5) is placed at the area of interest. The light emitted by the device interacts with the skin tissues, and the resulting reflected (remitted) portion is received by the device. This spectral information is converted into a color image. This image then goes through a supervised classification process based on the application of regression analysis techniques [221], which are mathematical tools extensively employed in analytical investigations involving skin color [120, 223]. In the training phase of this process, a set of simulated colors is obtained by running the model using different combinations of parameters. In the classification phase, for each color in the real image, the closest match is found in the training set. It is assumed that each color predicted by the model corresponds to a specific set of histological parameters [80], and for each derived parameter, a parametric map is created showing the magnitude of this parameter at each pixel location. It is important to note that similar colors may result from different combinations of parameters, a phenomenon known as

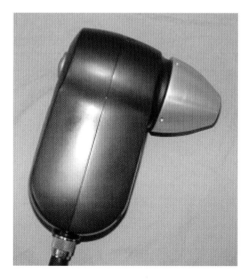

FIGURE 5.5

Hand-held device used in spectrophotometric intracutaneous analysis (SIAscopy) [103]. Reprinted from Journal of Plastic, Reconstructive & Aesthetic Surgery, 60, Govidan, K., Smith, J., Knowles, L., Harvey, A., Townsend, P., and Kenealy, J., Assessment of nurse-led screening of pigmented lesions using SIAscope, 639–645, © 2007, with permission from Elsevier.

metamerism [135]. For example, the presence of melanin in the dermis is a sign of melanoma, and it has a characteristic hue (Section 4.4.1). However, in specimens with thin papillary dermis, certain concentrations of melanin in the epidermis may result in similar hues. Accordingly, the first procedure performed in the SIAscope framework is the assessment of the thickness (and the quantity of collagen) in the papillary dermis using infrared light [177]. Such assessment is based on the negligible absorption of light by melanin in this domain and on the assumption that infrared light is forward scattered in the reticular dermis [50]. Despite the fact that the outputs of the model have not been quantitatively compared with measured skin reflectance data, the disease-related features identified by the SIAscope framework, notably with respect to melanoma, present a relatively high sensitivity and specificity. The former measure indicates how sensitive a test is to the presence of a given medical condition, in this case melanoma. The latter indicates how specific (in comparison with other tests) a test is to a given medical condition. However, the usefulness of employing SIAscope in the diagnosis of pigmented skin lesions, such as melanoma, still remains to be fully assessed since independent studies on this issue have resulted in conflicting conclusions [103, 109].

K–M models cannot be considered thorough models of optical radiation transfer since they lack a more detailed analysis of the structure and optical properties of the different skin tissues. In spite of that, the recent extensions to the original two-flux theory have improved their accuracy and increased their applicability to tissue optics studies. To further enhance their efficacy, skin researchers should take advantage of the substantial advances in K–M modeling with respect to the simulation of light interaction with plant leaves for remote-sensing applications [23, 124]. We remark that the measurement and modeling of light distribution in organic tissues is central in several areas of biological research, from the investigation of plant photosynthesis to the study of human photomedicine [237].

5.3 DIFFUSION THEORY–BASED MODELS

Photon propagation in optically turbid media, such as skin tissues, can be described by the time and energy independent equation of radiative transport known as the Boltzmann photon transport equation [123]. This equation requires the optical properties of the medium to be expressed in terms of the scattering coefficient, absorption coefficient, and phase function. The diffusion theory (DT) can be seen as an approximate solution of this equation [266]. It assumes a scattering-dominated light transport, and it combines the scattering coefficient and the phase function in one parameter, called reduced scattering coefficient, which is given by

$$\mu_s{}' = \mu_s(1-g).$$ (5.1)

Models based on the diffusion approximation [114] or combined with other approaches, such as the K–M theory [265] or Monte Carlo (MC) methods [273], have been used in biomedical investigations involving light propagation in turbid media. Although such models could be employed in investigations involving human skin, which can be described as turbid media, more predictive solutions may be obtained using models that take into account specific skin biophysical characteristics. For example, Schmitt et al. [217] presented a multilayer model that describes the propagation of a photon flux in the epidermal, dermal, and subcutaneous tissue layers of skin and assumes that the specimen is illuminated by a collimated, finite aperture light source. However, models whose implementation is based on (volumetric) absorption and scattering coefficients, like the model proposed by Schmidt et al. [217], have their usefulness limited by the practical difficulty of obtaining accurate values for the scattering and absorption coefficients for the different skin tissues.

Farell et al. [83] proposed a model based on the diffusion theory to be used in the noninvasive determination of the absorption and scattering properties of mammalian tissues. Their model incorporates a photon dipole source approximation in order to satisfy the tissue boundary conditions (Figure 5.6), namely light being remitted from a tissue from a point different from the incidence point, and the presence of thin layers of dirt, blood, or other fluids on the surface of the tissue under investigation.

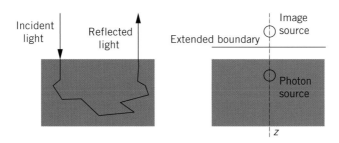

FIGURE 5.6

Sketch illustrating the boundary conditions taken into account by the dipole approximation. Left: light being remitted from a tissue from a point different from the incidence point. Right: extended boundary to account for the presence of thin layers of dirt, blood, or other fluids on the surface of the tissue under investigation.

This dipole approximation was originally used by Fretterd and Longini [96] and Hirko et al. [115], and further developed by Eason et al. [79]. Farrell and Patterson [83] considered two different boundary conditions: a refractive index mismatch, such as at a tissue and air interface, and a refractive index–matched interface. The predictions of their model were compared with skin optical properties obtained using MC simulations, and the results presented a limited agreement.

Doornbos et al. [72] proposed a hybrid method, based on the diffusion theory, for measuring optical properties and deriving chromophore concentrations from diffuse reflection measurements at the surface of a turbid media. Their method consists in measuring tissue reflectance and using the diffusion approximation to obtain the optical properties from the values measured in the wavelength domain for which this approximation holds. Doornbos et al. [72] consider the 650–1030 nm domain in their experiments. Initially, they approximate the wavelength dependency of the reduced scattering coefficient using a phase function designed to follow the Mie scattering theory (Section 2.3.2) and use these values to recalculate the tissue absorption and scattering coefficients with a higher accuracy. These coefficients are used to determine the concentration of the absorbers, namely water, oxyhemoglobin, and deoxyhemoglobin, as well as oxygen saturation. Although this hybrid method has provided results in the physiological range (within the same order of magnitude of actual in-vivo values), the accuracy of the modeled in-vivo concentrations cannot be properly assessed due to the difficulties involved in the direct and simultaneous measurement of these concentrations in living tissue.

Models based on the diffusion theory are amenable to analytic manipulation, place minor constraints on the type of sample, and are relatively easy to use [205]. However, the diffusion approximation, also called the P1 approximation [209], is applicable only under certain conditions. First, the measurement point needs to be remote from the light source, i.e., at a distance from the surface where the incident light beam has become completely diffuse [117]. Second, the absorption coefficient of the medium needs to be much smaller than its reduced scattering coefficient [62]. In other words, this approximation can be successfully applied only when scattering events are more probable than absorption events. This is usually the case for mammalian tissues in the red and near infrared regions of the light spectrum [92]. Not surprisingly, diffusion theory–based models have been used in medical applications involving red lasers [266, 285]. When the absorption coefficient of a turbid medium is not significantly smaller than the scattering coefficient, the diffusion theory provides a poor approximation for the photon transport equation [43, 216, 238, 285]. Other approximations for the transport equation exist. For example, the P3 approximation [236] can provide more accurate results near boundaries and sources [62], while the Grosjean approximation [14], also known as modified diffusion theory, is as effective as the P3 approximation but less computationally intensive [63]. However, the application of these approximations have been limited to internal tissues, tumors, and phantoms. Phantoms are objects resembling organic materials in mass, composition, and dimensions, which are used in biomedical investigations on the absorption of radiation in living tissues [18, 206].

5.4 RADIATIVE TRANSPORT MODELS

The K–M and diffusion theories mentioned in the previous sections can be seen as special cases of radiative transfer phenomena. When deterministic accurate solutions of the radiative transport equation in biological tissues are required, more robust methods need to be used, e.g., the successive scattering technique, Ambartsumian's method, the discrete ordinate method, Chandrasekhar's X and Y functions, and the adding-doubling method [201]. Their applicability, however, is usually limited to simple conditions and slab (an infinite plane parallel layer of finite thickness [201]) geometries. A comprehensive review of these methods is beyond the scope of this book, and the interested reader is referred to the texts by van de Hulst [260] and Prahl [201]. In this section, we highlight applications involving two of these methods in investigations of light interaction with human skin, namely the adding-doubling method and the discrete ordinate method.

The adding-doubling method has several advantages with respect to the other radiative transfer methods mentioned above. It permits asymmetric scattering, arbitrarily thick samples, Fresnel boundary conditions, and relatively fast computation [201]. The adding method requires that the reflectance and transmittance of two slabs be known. They are used to compute the reflectance and transmittance of another slab comprising these two individual slabs. Once the transmittance and reflectance for a thin slab are known, the reflectance and transmittance for a target slab can be computed by doubling the thickness of the thin slab until it matches the thickness of the target slab (Figure 5.7). In the original definition of this doubling method, it is assumed that both slabs are identical [260]. Later on, this method was extended to include the addition of two nonidentical slabs [201].

Prahl et al. [205] applied an inverse adding-doubling (IAD) method ("inverse" implying its use as an inversion procedure) to determine optical properties, namely scattering coefficient, absorption coefficient, and asymmetry factor, of biological tissues. The IAD is an iterative method which consists of guessing a set of optical properties, calculating the reflectance and

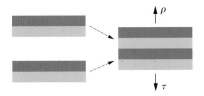

FIGURE 5.7

Sketch illustrating the application of the adding-doubling method to compute the reflectance and transmittance of a target tissue slab.

transmittance using adding-doubling method, comparing the calculated values with the measured reflectance and transmittance, and repeating the process until a match is obtained. This method may be used when the propagation of light through the specimen can be described by the one-dimensional radiative transport equation. Its accuracy, however, depends on the criteria applied to define a "sufficiently thin slab" [201]. There are also restrictions on the sample geometry; i.e., it must be an uniformly illuminated and homogeneous slab [205]. The IAD method has also been used to process spectral data (reflectance and transmittance) measured with a spectrophotometer equipped with an integrating sphere in order to derive (in-vivo and in-vitro) optical properties of skin and subcutaneous tissues [28, 255]. For this type of application, the main drawback of this method is the possibility of scattering radiation losses at the lateral sides of the sample [205]. Such radiation losses can erroneously increase the value computed for the optical properties [28].

Nielsen et al. [182] have proposed a skin model composed of five epidermal layers of equal thickness, a dermal layer, and a subcutaneous layer. The radiative transfer equation associated with this layered model is solved using the discrete ordinate algorithm proposed by Stamnes et al. [234] for the simulation of radiative transfer in layered media. Incidentally, Stam [232] has previously applied a similar approach in his model aimed at image synthesis applications (Section 6.2). The discrete ordinate method divides the radiative transport equation into n discrete fluxes to obtain n equations with n unknowns. These equations are solved using numerical techniques. Numerical linear algebra packages, such as EISPACK [53] and LINPACK [67], are usually used for that purpose [234]. This method is feasible when the phase function can be expressed as a sum of Legendre polynomials [41]. For highly asymmetric phase functions, it is necessary to consider a large number of fluxes, which may result in a numerically ill-conditioned system of equations [201]. The subdivision of the epidermis into five layers allow Nielsen et al. [182] to simulate different contents and size distributions of melanosomes (Figure 5.8). Their model accounts for the absorption associated with the

  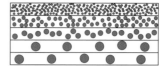

FIGURE 5.8

Diagram depicting three cases of melanin distribution in the five layers used by Nielsen et al. [182, 183] to represent the epidermis. Left: melanin is equally distributed in all layers. Middle: melanin is found in the melanosome particles located in the basal layer of the epidermis. Right: melanosome particles are distributed with varying sizes throughout the epidermis.

presence of blood [202] and melanosomes [129]. The absorption of keratin in the ultraviolet region [31] is also taken into account. In addition, the model accounts for the scattering by small particles using the Rayleigh phase function and for the scattering by large particles using the HGPF. The latter was selected due to the limited knowledge about the actual phase functions for these turbid media and for mathematical tractability since it can be expanded in terms of Legendre coefficients, which are employed in the discrete ordinate formulation. The reflectance curves obtained using their model showed good qualitative agreement with measured curves. They were also able to reproduce counter-intuitive empirical observations [82]. These observations indicated a higher reflectance at wavelengths below 300 nm for individuals with higher level of pigmentation as opposed to a lower reflectance for individuals with a lower lever of pigmentation. Other experimental investigations by Kölmel et al. [145], however, found a different relationship, i.e., a gradual decrease in reflectance with increasing pigmentation. According to Nielsen et al. [182], this apparent discrepancy may be explained by a putative shorter post-tanning period considered in the experiments by Kölmel et al. [145], which, in turn, may not have allowed for a fragmentation of the melanosomes. Nielsen et al. [183] later employed their model in the investigation of the role of melanin depth distribution in photobiological processes associated with the harmful effects of ultraviolet radiation on human skin. In this investigation, they employed a coupled atmosphere tissue–discrete ordinate radiative transfer (CAT-DISTORT) model [134] to account for solar light incidence as well as atmospheric and physiological conditions.

5.5 MONTE CARLO–BASED MODELS

The models reviewed in the previous sections were based on deterministic approaches. In this section, we will discuss models and applications based on a stochastic simulation approach introduced by Wilson and Adam [276] into the domain of biomedical optics. This approach consists in the application of MC methods to simulate the propagation and absorption of light in organic tissues (Section 3.2.2). We remark that, due to the experimental data scarcity, deterministic and stochastic modeling frameworks usually rely on mathematical functions to describe the bulk scattering of the material under investigation. In MC simulations, these functions are used to determine the direction of light after a scattering event. Clearly, the selection of an inappropriate function to approximate the bulk scattering of a material may introduce significant errors in the simulation process. For this reason, we start this section by taking a closer look at one the functions often employed in the modeling of light and skin interactions.

In 1984, Bruls and van der Leun [38] suggested that their measurements of the (bulk) scattering profile of stratum corneum and epidermis tissues (Section 4.3) could be approximated by one of the single particle phase function tabulated by van de Hulst [260, 261], namely the HGPF. Jacques et al. [128] followed Bruls and van der Leun's suggestion and attempted to approximate the measured scattering profile of another skin tissue, namely dermis, using the HGPF with an asymmetry factor $g = 0.81$. Yoon et al. [286] used similar asymmetry factor values for their studies involving human aorta. These investigations involving dermis and aorta tissues were aimed at specific medical applications and conducted with a HeNe laser (632.8 nm). Incidentally, at that time, Jacques et al. [128] appropriately stated that "the use of the HG function to specify radiant intensity for thicker samples is only descriptive, and should be distinguished from the customary use of the HG to described single particle phase function."

In 1988, Prahl [201] proposed a MC simulation framework for light transport in tissue during laser irradiation which took into account the HGPF approximations. In order to compute the trajectories of the scattered photons, Prahl [201] used a warping function provided by Witt [277], which was derived from the HGPF (Γ_{HG}) by setting

$$\xi_1 = 2\pi \int_{-1}^{\cos\alpha} \Gamma_{HG}(\cos\alpha', g) \, d\cos\alpha', \tag{5.2}$$

and finding upon integration that

$$\cos\alpha = \frac{1}{2g} \left\{ 1 + g^2 - \left[\frac{1 - g^2}{1 - g + 2g\xi_1} \right]^2 \right\}, \tag{5.3}$$

where ξ_1 is an uniformly distributed random number on the interval $[0, 1]$. For symmetric scattering ($g = 0$), the expression $\cos\alpha = 2\xi_1 - 1$ should be used [204]. Since an azimuthal symmetry of the phase function is assumed, the azimuthal angle can be generated using $\beta = 2\pi\xi_2$, where ξ_2 is a random number uniformly distributed on the interval $[0, 1]$.

In 1989, van Gemert et al. [264] attempted to fit the HGPF to the goniometric measurements of Bruls and van der Leun [38] using the least squares method to determine suitable values for the asymmetry factor g (Figures 5.9 and 5.10). Although other numerical techniques, such as the application of root mean square (RMS) error metric (Figures 5.9 and 5.10), may result in values for g that can provide a closer quantitative agreement with the data measured by Bruls and van der Leun, the HGPF approximation may not provide the degree of approximation required for applications demanding a higher level of accuracy. Furthermore, it has been demonstrated that a generalized use of the HGPF may negatively affect the accuracy and predictability of

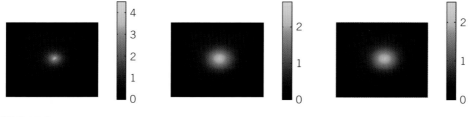

FIGURE 5.9

Comparison of measured and modeled scattering diagrams (orthographic projections) for an epidermis sample considering light incident at 436 nm. Left: using measured data [38]. Middle: using the HGPF with $g = 0.761$ (obtained using the RMS error metric [21]). Right: using the HGPF with $g = 0.748$ (obtained using the least squares method [264]).

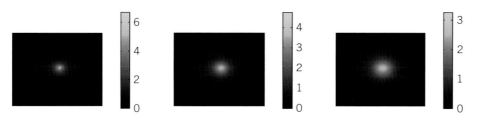

FIGURE 5.10

Comparison of measured and modeled scattering diagrams (orthographic projections) for an epidermis sample considering light incident at 546 nm. Left: using measured data [38]. Middle: using the HGPF with $g = 0.821$ (obtained using the RMS error metric [21]). Right: using the HGPF with $g = 0.781$ (obtained using the least squares method [264]).

light transport simulations [20, 21, 74, 178]. Nevertheless, the HGPF is often employed to approximate the bulk scattering of skin tissues due to its convenient mathematical tractability. Recall that this function is neither based on the mechanistic theory of scattering [128] nor was it originally devised to approximate the bulk scattering of organic tissues [113], i.e., the asymmetry factor g has no direct connection with the underlying biophysical phenomena involved in the interactions of light with these tissues. Although its use may be justifiable in the absence of measured scattering data or more accurate approximations, when such data is available, a more effective option is to use the data directly (e.g., through simple table lookups [20, 21]).

MC-based models, used in skin-related applications in biomedicine, colorimetry, and pattern recognition [47, 170, 172, 179, 204, 229, 253] usually provide only reflectance and transmittance readings for skin samples, i.e., the bidirectional reflectance distribution function (BRDF) and bidirectional transmittance distribution function (BTDF) quantities for the whole skin are not computed. We remark that most of these models are aimed at laser applications, and comparisons of modeled reflectance and transmittance values with actual measured values are scarce. This latter aspect makes the assessment of their correctness difficult.

Besides their direct use in simulations of light and skin interactions, MC models are also employed in other types of applications such as the evaluation of modeling frameworks based on deterministic approaches and the determination of skin optical properties and other biophysical attributes through indirect (inverse) procedures. Some works involving these applications have been outlined earlier. In the remainder of this section, we provide further examples of these practical contributions of MC models to skin optics investigations.

Shimada et al. [224] proposed a regression analysis algorithm to determine melanin and blood concentration in human skin. In their investigation, they applied the modified Beer–Lambert law and considered three-layered (epidermis, dermis, and subcutaneous tissue) skin phantoms. To assess the accuracy of their predictions, they employed a general purpose MC algorithm for light transport in multilayered tissues (MCML) developed by Wang et al. [274]. The same model was employed by Nishidateet al. [186] in their regression analysis investigation aimed at the estimation of melanin and blood concentration in the human skin as well as oxygen saturation. Although their algorithm was also based on modified Beer–Lambert law, it considered two-layered (epidermis and dermis) skin phantoms and employed the MCML model not only to verify the fidelity of their predictions but also to derive input data (conversion vectors) from a number of MCML-simulated absorption spectra. These vectors were used to compensate for the nonlinearity between the regression coefficients and the melanin and blood concentrations.

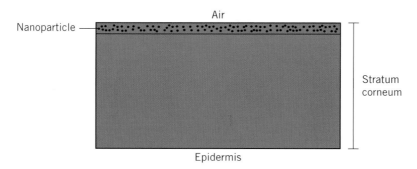

FIGURE 5.11

Sketch illustrating the two-layered stratum corneum model employed by Popov et al. [199] to investigate the optimal size for TiO_2 nanoparticles to be used as a UV-B skin-protective compound in sunscreens.

Popov et al. [199] employed a two-layer model for the stratum corneum (Figure 5.11) to investigate the optimal size for TiO_2 nanoparticles to be used as an UV-B skin-protective compound in sunscreens. The spherical nanoparticles are embedded in the top sublayer, which is assumed to be thinner than the bottom sublayer (with a ratio of 1:20). MC techniques [198] are then used to simulate the propagation of photons in both sublayers. The HGPF is employed to describe the scattering of the stratum corneum matrix, and a phase function derived from the Mie theory is used to describe light scattering from the nanoparticles. As a result, a hybrid scattering function formed by a weighted average of a Mie phase function and the HGPF is used to simulate scattering in the stratum corneum top sublayer. They also used as a parameter the volumetric absorption coefficient of the stratum corneum. Their simulations consisted in varying the concentration (0–1%) and diameter (25, 62, 85, 125, 150, and 200 nm) of the particles to determine which combination could be most effective to prevent light penetration in the stratum corneum. As expected, the penetration is reduced by increasing the volume concentration of the particles. According to their simulations, which have not been compared with actual in-situ experiments [161, 230], the particles with diameter equal to 62 nm provide the most effective barrier to light penetration.

Urso et al. [258] proposed different phantoms to investigate skin and cutaneous lesions. Initially, they prepared a multilayered skin-like phantom and another multilayered skin-like phantom with an embedded melanoma-like phantom inside (Figure 5.12). In order to characterize the optical features (absorption coefficient, scattering coefficient, and asymmetry factor) of the melanin and synthetic blood used in their phantoms, they employed an inversion procedure in which a computer multilayered skin model was used to

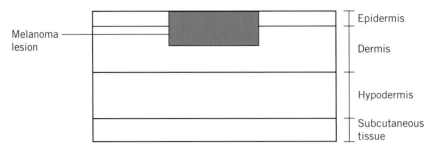

FIGURE 5.12

Sketch illustrating a multilayered skin-lesion phantom proposed by Urso et al. [258] to be used in the investigation of skin and cutaneous lesions.

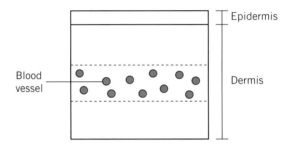

FIGURE 5.13

Diagram depicting the two-layered skin model proposed by Shi and DiMarzio [222].

simulate their phantoms. They then ran a series of MC simulations using the general purpose MCML algorithm [274]. They then compared their simulated diffuse reflectance and transmittance values with measured diffuse reflectance and transmittance values available in the literature [116] and adjusted the parameters accordingly. The process was repeated until an error equal to or less than 10% was obtained.

Shi and DiMarzio [222] used a regression analysis approach based on the modified Beer–Lambert law, similar to the one used by Nishidate et al. [186], to determine the concentration of skin chromophores. Their approach, however, employed an inhomogeneous two-layered (epidermis and dermis) skin model based on the model proposed by Kobayashi et al. [142]. The distinct feature of the skin model proposed by Shi and DiMarzio [222] is the presence of blood vessels represented by infinite cylinders, which are randomly distributed in a blood sublayer with an uniform dermal background (Figure 5.13). Furthermore, in their investigation, the MC simulations are used to generate the training sets used in the regression analysis process. In order to demonstrate the correctness of their framework, they ran MC simulations that use optical features as input parameters (chromophore concentrations and blood vessel geometry), which they extracted from skin spectral measured data using their framework. The resulting modeled absorption spectra closely matched the measured absorption spectra.

Most MC models share a similar mathematical formulation. The main factors that distinguish one model from another and affect their predictability are the level of abstraction used to describe the material under study and their parameter space. However, such details are usually not readily available in technical publications describing investigations that use MC models as simulation tools or as a reference for the evaluation of other models. This omission hinders the full assessment of the scientific contributions made by these investigations. We further address this issue, which is also relevant for computer graphics applications, in Chapter 9.

Biophysically inspired approach

6

The rendering of realistic images has always been one of the focal points of computer graphics. To achieve realism, it is necessary to model the appearance of different materials, which essentially involves the simulation of local light interactions that affect the spectral and spatial distribution of the light propagated by the material (Chapter 3). After all, as appropriately stated by Fournier [95], the only way that light acts at the rendering level is locally. Accordingly, a wide variety of local lighting models have been proposed in the computer graphics literature. Initially, these models used the approach "one size fits all"; i.e., models were designed to be applied to many different materials. Slowly, the model design became more specialized, especially with respect to organic materials. Model designers started to look more closely to material appearance attributes in order to obtain believable results. We call this approach biophysically inspired (or biophysically motivated) as opposed to biophysically based. While biophysically based models attempt to accurately simulate fundamental photobiophysical processes that determine the appearance of natural materials, biophysically inspired models focus on the generation of convincing images of these materials.

In this chapter, we will examine two biophysically inspired models that have been used to generate believable images of human skin. Although both models have been designed to have a general purpose, i.e., to assist the rendering of different materials, they are discussed in this book in the context of skin appearance modeling. The first model, the multiple-layer scattering model, employs a stochastic approach, while the second, the discrete ordinate model, relies on a deterministic approach. These models, however, share some technical similarities such as the use of phase functions to approximate the materials' bulk scattering. Although both models employ techniques developed in other scientific domains, they have the merit to have introduced

DOI: 10.1016/B978-0-12-375093-8.00006-X

relevant tissue optics concepts and terminology to the computer graphics community. More importantly, these works brought the appearance modeling of organic materials, especially human skin, to the forefront of realistic rendering research.

6.1 THE MULTIPLE-LAYER SCATTERING MODEL

In 1993, Hanrahan and Krueger [110] proposed a model to simulate subsurface reflection and transmission from layered surfaces [100]. This intuitive idea of a layered surface model has appeared before in fields such as remote sensing [23, 124] and tissue optics (Chapter 5). The model proposed by Hanrahan and Krueger (H-K model) incorporates tissue optics concepts and techniques. It can be used to simulate the scattering profile of organic (e.g., skin and leaves) and inorganic materials (e.g., snow and sand). In the case of human skin, it was modeled by Hanrahan and Krueger as having two layers, namely epidermis and dermis.

6.1.1 Overview

Hanrahan and Krueger [110] assumed planar surfaces and used Fresnel coefficients to find how much light will pass through the outermost surface of the material. The model then evaluates the amount of scattering and absorption within each layer, including the reflection and transmission effects at each internal boundary. The bidirectional reflectance distribution function (BRDF) and bidirectional transmittance distribution function (BTDF) are described by a combination of the reflection function on the outer surface and the internal subsurface scattering handled by a Monte Carlo algorithm, which was originally proposed by Prahl [201, 204] to investigate laser irradiation in tissue (Section 5.5).

   In the H-K model, it is assumed that if a material is a mixture of several materials, then the mixture can be considered to be a uniform and homogeneous combination whose properties are given by a sum of the descriptors of the components weighted by percentages. The material descriptors include the index of refraction, the absorption cross section, the scattering cross section, the depth (or thickness), and a phase function (the Henyey–Greenstein phase function [HGPF]). The absorption and scattering cross sections used by Hanrahan and Krueger correspond to the volumetric absorption and scattering coefficients, respectively (Section 2.5). In this chapter, for the sake of consistency with the tissue optics literature, we use the terms absorption coefficient and scattering coefficient instead of the terms absorption cross section and scattering cross section used by Hanrahan and Krueger [110].

6.1.2 Scattering simulation

The H-K model assumes that the reflected radiance from a surface has two components (Figure 6.1). One arises due to surface reflectance (L_{rs}) and the other due to subsurface volume scattering (L_{rv}). It also assumes that the transmitted radiance has two components (Figure 6.1). One, called *reduced intensity*, represents the amount of light transmitted through the layer without scattering inside the layers, but accounting for absorption (L_{ri}). The other represents the amount of light scattered in the volume (L_{tv}). Similarly, the BRDF and BTDF also have two components, and the relative contributions of the surface and subsurface terms are modulated by the Fresnel coefficients. Clearly, the variations on the polar angle of incidence, given by θ_i, will affect the value of these coefficients, which, in turn, will affect the magnitude of BRDF and BTDF components. Since there is no dependency on the azimuthal angle of incidence, the H-K model can be classified as an isotropic model.

Surface reflection

As indicated by Hanrahan and Krueger [110], surface reflection is handled using the Torrance and Sparrow model [249]. In this model, reflected energy is attenuated due to orientations of surface microfacets according to a multiplicative exponential factor $e^{-\varsigma^2/v^2}$, where the angle ς corresponds to the inclination of the microfacets with respect to the normal of the mean surface,

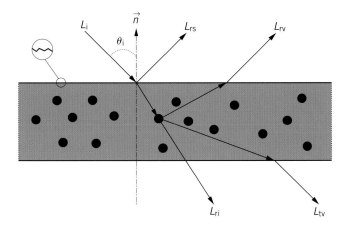

FIGURE 6.1

Sketch of the scattering geometry used in the H-K model.

and the parameter v can be seen as their root mean square (RMS) slope. It should be cautioned that the value of the parameter v should not be very large, otherwise the results may become physically implausible. According to experiments performed by Torrance and Sparrow [249], the value of this parameter should range from 5 to 100 for inorganic materials such as ground glass surfaces. Other issues related to the use of this model in conjunction with the H-K model are addressed in Section 6.1.3.

Subsurface reflection and transmission

Hanrahan and Krueger [110] examined the application of first-order approximations for the analytical solution of the integral transport equation assuming only a single scattering event [41]. The first-order solution for L_{tv} is obtained by replacing L_{ri} in the integral equation [41], and the first-order solution for L_{rv} is obtained by applying Seeliger's formulation for diffuse reflection. Seeliger [220] attempted to explain experimental deviations from Lambert's law by relating the scattering to the structure of a particular material under consideration. He quantitatively used a hypothesis originally proposed by Bouguer [37], which consists in assuming a surface to be composed of countless small elementary mirrors disposed at all possible angles [26, 33].

These first-order solutions are then used by Hanrahan and Krueger in the first step of a refinement approach, which consists in substituting the ith-order solutions in the integral equation and solving to get the $(i + 1)th$-order solutions. As stated by Hanrahan and Krueger, this analytical refinement approach "quickly becomes intractable." Thus, alternatively, the H-K model applies an algorithmic approach for computing light transport in a layered semifinite turbid media with different albedos (γ). More precisely, the subsurface scattering is simulated using a Monte Carlo algorithm previously applied in the biomedical optics field (Section 5.5). This algorithm can be concisely described as follows:

1. As a ray enters the layer at the origin, initialize point o as the origin and a vector $\vec{s}$ as the direction at which the ray enters the layer. Set the ray weight to $w = 1$.

2. Repeat the following steps until the ray exits the layer or its weight drops below a given threshold.
   (a) Estimate the distance to the next iteration using the following formula:

$$\Delta_s = -\frac{\ln \xi_1}{\mu_s + \mu_a},$$

where ξ_1 corresponds to an uniformly distributed random number $\in [0,1]$ and μ_s and μ_a correspond to the volumetric scattering and absorption coefficients, respectively.

(b) Compute the new position, i.e., $o \leftarrow o + \Delta_s \vec{s}$.

(c) Set the ray weight to $w \leftarrow w\,\gamma$, where γ corresponds to the albedo.

(d) Compute the cosine of the scattering angle using Equation 5.3.

(e) Perturb the ray direction using

$$\vec{s} = \vec{s}\,\cos\alpha + \vec{\imath}\,\sin\alpha,$$

where

$$\vec{\imath} = \begin{pmatrix} (\vec{s}.x\,\cos\beta\,\cos\Theta - \vec{s}.y\,\sin\beta)\sin\Theta \\ (\vec{s}.y\,\cos\beta\,\cos\Theta + \vec{s}.x\,\sin\beta)\sin\Theta \\ \sin\alpha \end{pmatrix},$$

and

$$\beta = 2\pi\xi_2,$$

in which ξ_2 corresponds to an uniformly distributed random number $\in [0,1]$ and $\cos\Theta = \vec{s}.z$ and $\sin\theta = \sqrt{1 - (\vec{s}.z)^2}$.

3. Divide the sphere surrounding the materials into regions of equal solid angle increments. These regions are considered to be bins receiving emitted rays. When a ray exits the material, add the weight of the ray to the weight of the bin that receives it.

6.1.3 Implementation issues

The H-K model was originally implemented as an extension to a standard ray tracer developed using *Rayshade* [143], a public domain software for rendering applications written in C, *lex,* and *yacc.*

The skin tissues, namely epidermis and dermis, and their constituents are considered as dielectrics, and Hanrahan and Krueger [110] assign to them indices of refraction between 1.37 and 1.5. Although Hanrahan and Kruger do not provide a direct reference to the source from which the values for the absorption and scattering coefficients and asymmetry factors used in their experiments were obtained, it was implied that they could be obtained from the work by van Gemert et al. [264].

In the H-K model, a ray enters the material and is repeatedly propagated from one scattering event to the next. Each scattering event attenuates the weight associated with the ray by a fixed factor which corresponds to the albedo. If the ray is not absorbed, it will eventually be scattered out of the material. In this last scattering event, the ray may leave the layer. In this case, according to Hanrahan and Krueger [110], the weight should be adjusted using the distance to the boundary. Although Hanrahan and Krueger did not explain how this adjustment should be performed, the final result may not be significantly affected by the weight adjustment with respect to the last interaction.

The procedure to be adopted when $\sin \Theta = 0$ was also left to the reader's interpretation perhaps for similar reasons. In this case, one could simply discard the current $\vec{s}$ for which $\sin \Theta = 0$, and compute another one. Since this situation is not likely to occur often, the possible implications in the final result would probably be negligible as well.

In the original paper [110], the warping function used to compute the cosine of the scattering angle has the term $|2g|$ in the denominator instead of $2g$, which effectively takes away the possibility of backscattering. This is quite acceptable for organic materials, which are characterized by forward scattering. If the parameter g is set to zero, the warping function reduces to $2\xi - 1$, where ξ corresponds to an uniformly distributed random number $\in [0, 1]$.

Hanrahan and Krueger did not provide details in their paper on how the Torrance–Sparrow model [249] was used to account for surface reflection on the outermost layers. However, since their framework is based on a stochastic ray tracing approach, one may consider that the Gaussian distribution of microfacets adopted in the Torrance–Sparrow model was used to obtain a warping function to perturb the reflected rays.

6.1.4 Strengths and limitations

The H-K model has the merit of being one of the first computer graphics models to address important issues related to the simulation of light interaction with biological materials. However, because of its generality, it tends to overlook important specific characteristics and properties of organic materials, such as the mechanisms of absorption of light by natural pigments and their specific absorption coefficients. In the H-K model, these aspects are considered only implicitly through the use of coefficients available in the biomedical literature, which, in turn, were obtained using inversion procedures. Hence, the reflectance and transmittance of skin specimens are not computed directly, but implicitly introduced into the model as the albedo.

In other words, the H-K model has to be considered as a scattering model, instead of a reflectance model, since reflectance and transmittance values are not computed.

The H-K model is isotropic, i.e., it only considers the polar angle of the light incident on the material surface. As described earlier, Hanrahan and Krueger indicated that their model can be combined with the Torrance-Sparrow model [249] in order to take into account the surface reflection on the outermost layers. The latter was, however, designed based on experimental data for inorganic materials, and its parameters are not biologically meaningful. Thus, it is not clear what criteria should be used in the selection of its parameters in order to model light interaction with organic materials.

The use of the HGPF in the subsurface scattering simulation of skin tissues raises some issues as well. For example, its main parameter, the asymmetry factor, has no direct connection with the underlying biophysical phenomena (Section 5.5). Furthermore, recall that the HGPF was initially meant to be used in tissue optics just as a function to fit multiple scattering data of skin measured at specific wavelengths by Bruls and van der Leun [38]. As discussed in Section 5.5, the HGPF approximations may deviate from the measured data considerably, and its generalized application to any organic tissue at any wavelength may lead to incorrect results, especially using asymmetry factors determined by fitting the HGPF to specific data sets that may have no relationship with the material at hand.

The evaluation of the H-K model did not include comparisons of model predictions with actual measured data, and it was based only in the visual inspection of computer-generated images. More specifically, Hanrahan and Krueger provided images illustrating the application of different modeling features and compared images rendered using the H-K model with images rendered using a Lambertian model (Figure 6.2).

Although the H-K model cannot be considered a predictive model, it can be used to render believable images of human skin, and it raised the bar for computer graphics research involving organic materials.

6.1.5 Extensions

In 2001, Ng and Li [180] proposed an extension to the H-K model which consists in adding a sebum (human oily secretion) layer on top of the epidermis layer. The work of Ng and Li has the merit of providing comparisons with actual measured data.

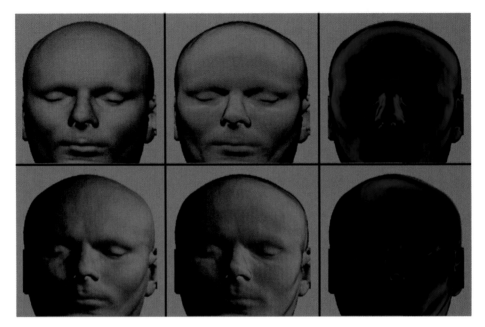

FIGURE 6.2

Comparisons of images generated using a Lambertian model with images generated using the
H-K model [110] and considering two angles of incidence, namely 0° (top row) and 45° (bottom
row). Left column: Lambertian model. Middle column: H-K model. Right column: relative
difference of both models, with the red color indicating more reflection from the new model and
the blue color indicating the opposite. Hanrahan, P., and Krueger, W. "Reflection from layered
surfaces due to subsurface scattering" SIGGRAPH, *Annual Conference Series* (August 1993),
pp. 165–174. © 2010 Association for Computing Machinery, Inc. Reprinted by permission.

6.2 THE DISCRETE-ORDINATE MODEL

The discrete-ordinate (DO) model proposed by Stam [232] was developed to
simulate the scattering behavior of skin. Spectral quantities related to skin
reflectance and transmittance are introduced into the model as input parameters, i.e., mechanisms of light absorption by natural pigments were only
implicitly considered. Although this biologically motivated model adopts a
simplified representation for skin structure and optical properties, its deterministic algorithms for the simulation of light transport can represent a viable
alternative for rendering applications that demand high interactive rates.

6.2.1 Overview

In the DO model formulation, skin is represented by a single layer with
constant optical properties and an uniform index of refraction. In addition,

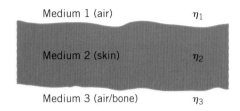

Medium 1 (air) η_1

Medium 2 (skin) η_2

Medium 3 (air/bone) η_3

FIGURE 6.3

Sketch describing the skin representation used by the DO model, where the skin is represented by a layer with an uniform index of refraction (η_2) bounded by two media with uniform refractive indices (η_1 and η_3) as well.

this layer is bounded by media having uniform indices of refraction as well (Figure 6.3). Following Hanrahan and Krueger [110], the DO model represents the skin depth along the z-direction and assumes that the skin properties are uniform in each xy-plane, i.e., the skin is horizontally uniform. The parameters used to model the skin layer are the optical depth, the albedo, and the asymmetry factor of the phase function (HGPF). Each parameter used in the DO model is dimensionless and varies from zero to one.

The DO model also uses a parameter to account for the roughness of the surfaces bounding the skin layer. This parameter controls the spatial distribution of light propagated at these interfaces. Incidentally, the surface roughness is assumed to be isotropic, i.e., only the polar (elevation) angle of the light incident on the rough surfaces matters. Hence, the DO model can be classified as isotropic.

6.2.2 Scattering simulation

In order to model a skin layer bounded by rough surfaces, Stam extended the work by Stamnes and Conklin [233] for a skin layer bounded by a smooth surface. This work is based on the discrete-ordinate approximation of the radiative transfer equation [41]. The method of discrete-ordinates divides the radiative transport equation into n discrete fluxes to obtain n equations with n unknowns [201]. These equations were solved numerically by Stam [232] using Fourier transforms and eigenanalysis [99]. His approach was inspired by the work of Jin and Stammes [134].

Surface reflection

The DO model takes into account the reflection and refraction from the rough surfaces by extending the BRDF model proposed by Cook and Torrance [55] and following the work of van Ginneken et al. [267] in which the surfaces are assumed to have a normal distribution of heights. The reflection due to an ambient light source is modeled by integrating, over all incident directions, the BRDF due to scattering in the skin layer whose computation is concisely described in the next section.

Subsurface reflection and transmission

In the formulation used in DO model, the radiance within the skin is obtained by considering its variation in an infinitesimal cylinder according to the following equation:

$$-\varpi \frac{dL}{d\varrho} = -L + \frac{\gamma}{4\pi} \int_{4\pi} \Gamma(\vec{s},\vec{i})L(\varrho,\vec{s})d\vec{s}, \tag{6.1}$$

where L is the radiance, γ is the albedo, ϖ is the cosine of the polar scattering angle, ϱ is the optical depth, $\vec{i}$ is the incident vector, $\vec{s}$ is the scattering vector, and $\Gamma(\vec{s},\vec{i})$ is the phase function.

Equation 6.1 is solved taking into account boundary conditions that related the BRDF and BTDF values, represented by $f_{r_{ij}}$ and $f_{t_{ij}}$ respectively (Figure 6.4), with radiance values at the air and skin interfaces. Initially, the reflection and transmission operators, represented by $\mathcal{R}_{ij}$ and $\mathcal{T}_{ij}$, respectively, are defined as

$$\mathcal{R}_{ij}\{L\}(\varrho,\pm\vec{i}) = \int_{2\pi} f_{r_{ij}}(\mp\vec{s},\pm\vec{i})L(\varrho,\mp\vec{s})\xi\,d\vec{s} \tag{6.2}$$

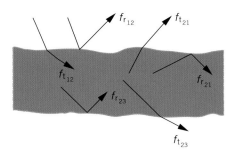

FIGURE 6.4

Sketch describing the BRDF and BTDF interactions, represented by $f_{r_{ij}}$ and $f_{t_{ij}}$, respectively, taken into account in the DO model.

and

$$\mathcal{T}_{ij}\{L\}(\varrho, \pm\vec{i}) = \int_{2\pi} f_{t_{ij}}(\mp\vec{s}, \pm\vec{i}) L(\varrho, \mp\vec{s}) \xi \, d\vec{s}. \qquad (6.3)$$

Then, the boundary condition at the skin surface ($\varrho = 0$) is given by

$$L(0, -\vec{i}) = f_{t_{12}}(-\vec{i}_0, -\vec{i}) + \mathcal{R}_{21} L(0, -\vec{i}), \qquad (6.4)$$

where $-\vec{i}_0$ is the incident vector on the air and skin interface.

The boundary condition at the bottom ($\varrho_b = 0$), assuming that there are no sources below the skin, is given by

$$L(\varrho_b, \vec{i}) = \mathcal{R}_{23}\{L\}(\varrho_b, \vec{i}). \qquad (6.5)$$

Finally, the BRDF due to subsurface scattering is given by

$$f_{rs}(-\vec{i}_0, \vec{v}) = \mathcal{T}_{21}\{L\}(0, \vec{i})/\varpi_0, \qquad (6.6)$$

where ϖ_0 is the cosine of the polar scattering angle with respect to $-\vec{v}_0$.

Similarly, the BTDF due to subsurface scattering is given by

$$f_{ts}(-\vec{v}_0, -\vec{v}) = \mathcal{T}_{23}\{L\}(\varrho_b, -\vec{v})/\xi_0, \qquad (6.7)$$

The skin BRDF and BTDF represented by Equations 6.6 and 6.7 are discretized with respect to a number of sample directions. This discretization process results in a collection of matrices that are precomputed for different values of the parameters used in the DO model. During this process, the phase function, namely the HGPF, is expanded into a cosine series whose coefficients are expressed in terms of Legendre functions [41, 260, 261]. The corresponding system of equations representing n discrete fluxes described by n equations with n unknowns is solved numerically using Fourier transforms and eigenanalysis [99].

6.2.3 Implementation issues

The precomputation of the matrices generates a large data set. In order to allow a practical use of this data in rendering frameworks, Stam [232] compressed it using an approximation based on cosine lobes. The cosine terms were chosen by visually comparing the data to the approximation. The data set was then further compressed by fitting a cubic Bèzier surface [93] to the data stored in the reflection and transmission matrices. The control vertices of the Bèzier surfaces were constrained to respect their symmetry, i.e., to obey the Helmholtz reciprocity rule (Section 2.6.1).

Stam [232] used the EISPACK [53] and LINPACK [67] numerical linear algebra packages to solve the system of equations. Alternatively, one could use their successor LAPACK [76]. Also, Stam made the DO model implementation available as a shader plugin for the rendering software package Maya [194].

6.2.4 Strengths and limitations

The DO model is only a scattering model since reflectance and transmittance quantities are not computed. Although it is biologically inspired, it does not take into account the structural characteristics of skin tissues and the biological processes that affect propagation and absorption of light in these tissues. The oversimplification of these biological processes, however, is not accompanied by the mathematical complexity of the algorithms used in the DO model. Although these are not as complex as the rare analytical solutions for radiative transfer problems found in the literature, they are certainly less straightforward than the algorithms used in Monte Carlo–based models. The main advantage of the DO model over stochastic approaches using Monte Carlo methods is speed, which is sustained using precomputation and compression schemes. We remark that the outputs of MC models can also be precomputed and compressed offline. We further address these alternatives for performance enhancement in Chapter 9.

On a rough surface, it is possible that some points are blocked from the light (shadowing effect) or from the view light (masking effect) [95]. The DO model takes into account these effects, which are rarely incorporated in computer graphics models of light interaction with organic materials. As mentioned above, in order to accomplish that, Stam extended the Cook and Torrance model [55] using a shadowing function proposed by van Ginneken et al. [267]. Both models were aimed at inorganic or man-made materials; however, and their applicability to biological materials has not been verified.

The discrete-ordinates approach to solve radiative transfer problems is suitable when the material phase function can be expressed as a sum of a few terms [201]. The HGPF used in the DO model can be expanded in a cosine series whose coefficients are expressed in terms of Legendre functions [41]. This contributes to relatively quick solutions for the radiative transfer equations. However, as pointed out earlier (Sections 5.5 and 6.1.4), the use of the HGPF in tissue optics is questionable in terms of its effects on the accuracy and predictability of the simulations.

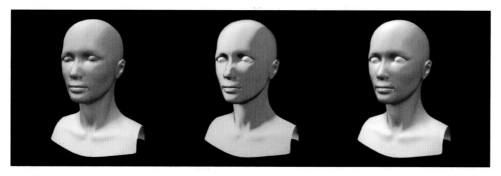

FIGURE 6.5

Images generated using the DO model (left), a Lambertian model (middle), and the H-K model (right) [232].

The original description of the DO model [232] does not include any comparison with measured data, only qualitative comparisons with results provided by other models. More specifically, an image generated using the model was compared to images generated using the H-K model and a Lambertian model (Figure 6.5). We remark, however, that an accurate assessment of a model benefits and costs ratio cannot be obtained solely through the visual inspection of rendered images. For example, one's perception of realism can be significantly altered by the use of different geometrical models and texture maps.

First principles approach

Different levels of abstraction can be used in the design of biophysically based local lighting models. In theory, one would like to use the lowest possible level of abstraction so that even small perturbations in the biophysical processes can be accounted for. In practice, the appropriate abstraction level is usually tied to data availability. After all, a detailed model described by many parameters would not be useful if there is no reliable data to be assigned to these parameters. First principles models attempt to work on the lowest level of abstraction for which there is available data. For instance, instead of using a volumetric absorption coefficient for a given tissue, their formulation can incorporate specific absorption coefficients for the tissue constituents.

Many models used in computer graphics rely on spectral parameters, such as reflectance and transmittance, whose values are either arbitrarily set by the user or obtained from direct measurements or inversion procedures. There are measured reflectance curves for human skin available in the biomedical literature, but they are limited to a number of skin types and restricted to a narrow range of illuminating and viewing angles. Furthermore, measured transmittance curves for the skin organ as a whole are scarce. These aspects highlight the need to develop models of light interaction with human skin which can compute not only its scattering properties (usually given in terms of BRDF and BTDF), but also its spectral properties (usually given in terms of reflectance and transmittance). The biophysically based spectral model, henceforth referred to as BioSpec, was proposed by Krishnaswamy and Baranoski [147] to address this need. BioSpec was the first model to use a first principles approach for computing both spectral and scattering quantities for skin specimens. For this reason, its design and evaluation are closely examined in this chapter.

113

7.1 OVERVIEW

The BioSpec model uses ray optics and Monte Carlo techniques to simulate the processes of light propagation (surface reflection, subsurface reflection, and transmission) and absorption in the skin tissues. It considers the stratification of skin into four semi-infinite main layers: stratum corneum, epidermis, papillary dermis, and reticular dermis. The model parameter space includes the following: the refractive index and thickness of each layer, the refractive index and the diameter of collagen fibrils, the extinction coefficient, concentration, and volume fraction of the main chromophores present in the skin tissues (eumelanin, pheomelanin, oxyhemoglobin, deoxyhemoglobin, β-carotene, and bilirubin) and the aspect ratio of the stratum corneum folds.

The propagation of light in the skin tissues is simulated as a random walk process [100], whose states are associated with the following interfaces:

1. air $\Leftrightarrow$ stratum corneum;
2. stratum corneum $\Leftrightarrow$ epidermis;
3. epidermis $\Leftrightarrow$ papillary dermis;
4. papillary dermis $\Leftrightarrow$ reticular dermis;
5. reticular dermis $\Leftrightarrow$ hypodermis.

Once a ray hits the skin specimen at interface 1, it can be reflected back or refracted into the stratum corneum. From there, the ray can be reflected and refracted multiple times within the skin layers before it is either absorbed or propagated back to the environment through interface 1. For body areas characterized by the presence of the hypodermis, the BioSpec algorithmic formulation assumes total reflection at interface 5. For body areas where hypodermis is virtually absent, interactions at interface 5 are turned off.

In the random walk implemented by the BioSpec model (Figure 7.1), the transition probabilities are associated with the Fresnel coefficients computed at each interface. The termination probabilities of this random walk are associated with the mean free path length (p) computed when a ray travels in the skin layers.

7.2 SCATTERING SIMULATION

The BioSpec model takes into account the following three components of the BDF of a skin specimen: surface reflectance, subsurface reflectance,

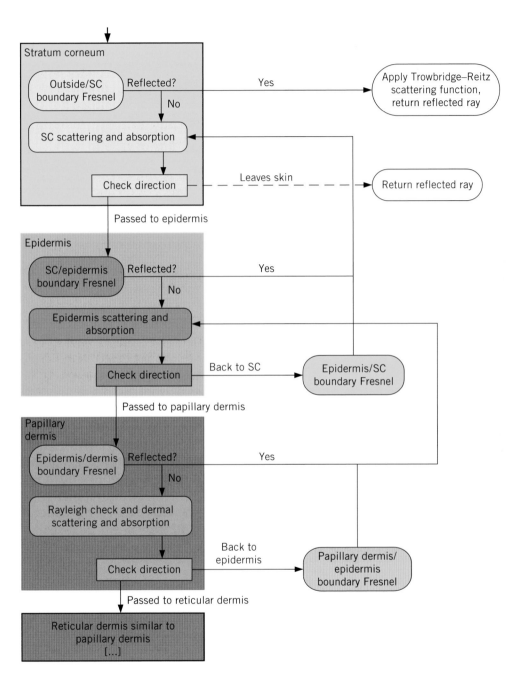

FIGURE 7.1

A flow chart describing the main stages of the random walk process simulated by the BioSpec model.

and transmittance. These components are affected by the refractive index differences at the interfaces, tissue scattering, and absorption of light by skin pigments. In the next sections, we describe how each of these components is simulated. Because of the stochastic nature of the simulations, several random numbers uniformly distributed in the interval [0, 1] and represented by ξ_i (for $i = 1..12$) are used in the BioSpec formulation.

After computing the Fresnel coefficient (F_R) at an interface, a random number ξ_1 is obtained. If $\xi_1 \leq F_R$, then a reflected ray is generated, otherwise a refracted ray is generated. The F_R is computed using the Fresnel equations and taking into account the refractive index differences of the media, namely stratum corneum, epidermis, papillary dermis, and reticular dermis, which are represented by η_{sc}, η_{ep}, η_{pd} and η_{rd}, respectively. The reflected ray is computed applying the law of reflection, and the refracted ray is computed applying Snell's law (Chapter 2).

7.2.1 Surface reflection

A portion of the light that interacts with the stratum corneum cells is reflected back to the environment following the computation of the Fresnel coefficients described earlier. The spatial distribution of the reflected light varies according to the aspect ratio of the stratum corneum folds (Section 4.3). In the BioSpec model, these mesostructures are represented as ellipsoids. The aspect ratio ($\sigma \in [0, 1]$) of these ellipsoids is defined as the quotient of the length of the vertical axis by the length of the horizontal axis, which are parallel and perpendicular to the specimen's normal respectively. As the folds become flatter (lower σ), the reflected light becomes less diffuse. In order to account for this effect, the reflected rays are perturbed using angular displacements obtained from the surface-structure function proposed by Trowbridge and Reitz [251], which represents rough air-material interfaces using microareas randomly curved. These displacements [146, 147] are given in terms of the polar perturbation angle

$$\theta_s = \arccos\left[\left(\left(\frac{\sigma^2}{\sqrt{\sigma^4 - \sigma^4\xi_2 + \xi_2}} - 1\right)b\right)^{\frac{1}{2}}\right], \tag{7.1}$$

where b is $\frac{1}{\sigma^2-1}$, and the azimuthal perturbation angle

$$\phi_s = 2\pi\xi_3. \tag{7.2}$$

7.2.2 Subsurface reflection and transmission

Scattering in either the stratum corneum or epidermis involves the perturbation of the incoming ray in both the polar (α_f) and azimuthal (β_f) angles. The scattering with respect to the azimuthal angle β_f is expected to be symmetric (equal in all directions) [204], thus $\beta_f = 2\xi_4\pi$ is used. The scattering direction with respect to the polar angle α_f is computed using a randomized table lookup algorithm. The polar scattering angles measured at a given wavelength by Bruls and Leun [38] (Section 4.3) are stored in a table, whose access indices correspond to the measured fractions of scattered radiation. For each ray a random number ξ_5 is generated, which is multiplied by the table size. The integer part of the resulting value is used to access the corresponding polar scattering angle stored in the table.

Every ray entering one of the dermal layers is initially tested for Rayleigh scattering (Section 2.3.2). If the test fails or the ray has already been bounced off one of the dermal interfaces, then the ray is randomized around the normal direction using a warping function based on a cosine distribution (Section 3.1.2), in which the polar perturbation angle, α_c, and the azimuthal perturbation angle, β_c are given by

$$(\alpha_c, \beta_c) = \left(\arccos\left((1 - \xi_6)^{\frac{1}{2}}\right), 2\pi\xi_7 \right). \tag{7.3}$$

In order to perform the Rayleigh scattering test, the spectral Rayleigh scattering amount, $\mathcal{S}(\lambda)$, is computed (Equation 7.6). This spectral quantity is associated with the probability that the Rayleigh scattering can occur [168]. A random number ξ_8 is then generated, and if $\xi_8 < 1 - e^{-\mathcal{S}(\lambda)}$, then the ray is scattered using polar (α_R) and azimuthal (β_R) perturbation angles. The perturbation angles are given by

$$(\alpha_R, \beta_R) = (\Theta_p, 2\pi\xi_9), \tag{7.4}$$

where the angle Θ_p is obtained using rejection sampling in conjunction with the Rayleigh phase function [168]

$$
\begin{aligned}
&\text{do} \\
&\qquad \Theta_p = \pi\xi_{10} \\
&\qquad \chi = \tfrac{3}{2}\xi_{11} \\
&\text{while} \quad (\chi > \tfrac{3}{4}(1 + \cos^2\Theta))
\end{aligned}
$$

Note that ξ_{10} and ξ_{11} must be regenerated during each iteration of the loop described earlier.

According to Jacques [126], collagen fibers occupy 21% of the dermal volume, and the Rayleigh scattering in this tissue can be approximated using spheres mimicking the ultrastructure associated with the random arrays of collagen fibrils of radius r_f. This results in a density of collagen fibers δ_f given by

$$\delta_f = 0.21 \left(\frac{4}{3} r_f^3 \pi \right)^{-1},$$ (7.5)

which is used in the BioSpec formulation to compute the spectral Rayleigh scattering amount [146, 168] through the following equation:

$$S(\lambda) = \frac{8\pi^3 \left(\left(\frac{\eta_f}{\eta_{de}} \right)^2 - 1 \right)^2}{3\delta_f \lambda^4} \left(\frac{h}{\cos\theta} \right),$$ (7.6)

where η_f is the index of refraction of the fibers, η_{de} is the index of refraction of the dermal medium, h is the thickness of the medium, and θ is the angle ($<90°$) between the ray and the normal of the specimen.

7.3 ABSORPTION SIMULATION

When a ray travels in a given layer, it is first scattered as described in the previous section. The ray is then tested for absorption. If the ray is not absorbed, then it is propagated to the next layer. The absorption testing done by the BioSpec model is based on Beer–Lambert law (Section 2.6.1). It is performed probabilistically every time a ray starts a run in a given layer. It consists of estimating the ray mean free path length (p) through the following expression:

$$p(\lambda) = -\frac{1}{\mu_{ai}(\lambda)} \ln(\xi_{12}) \cos\theta,$$ (7.7)

where $\mu_{ai}(\lambda)$ is the total absorption coefficient of the pigments of a given layer i and θ is the angle between the ray and the normal of the specimen.

If $p(\lambda)$ is greater than the thickness of the pigmented medium, then the ray is propagated, otherwise it is absorbed. The thickness of the stratum corneum, epidermis, papillary dermis, and reticular dermis are represented by h_{sc}, h_{ep}, h_{pd}, and h_{rd}, respectively.

The BioSpec model accounts for the presence of eumelanin, pheomelanin, oxyhemoglobin, deoxyhemoglobin, bilirubin, and β-carotene. The spectral molar absorption (extinction) coefficients for these pigments, denoted $\varepsilon_{eu}(\lambda)$, $\varepsilon_{ph}(\lambda)$, $\varepsilon_{o}(\lambda)$, $\varepsilon_{d}(\lambda)$, $\varepsilon_{bil}(\lambda)$, and $\varepsilon_{car}(\lambda)$, respectively, correspond to measured values (Figure 4.2). The total absorption coefficient for each layer is simply the sum of the absorption (extinction) coefficients of each pigment present in the layer. These coefficients are obtained by multiplying the molar absorption (extinction) coefficient of the pigment (which can be given either in $cm^{-1}(mol/L)^{-1}$ or $cm^{-1}(g/L)^{-1}$) by its estimated concentration in the layer (usually given in grams per liter).

It is difficult to accurately determine the baseline absorption coefficient for pigmentless skin tissues. Furthermore, because of its low magnitude [214] compared with the absorption coefficients of the skin chromophores, skin optics researchers usually assume that its effects are negligible [12]. For the sake of completeness, however, the baseline skin absorption coefficient, $\mu_{a_{base}}(\lambda)$, was included in the absorption equations used in the BioSpec model.

The stratum corneum total absorption coefficient is given by

$$\mu_{a_1}(\lambda) = \mu_{a_{base}}(\lambda) + \mu_{a_{cs}}(\lambda), \tag{7.8}$$

where $\mu_{a_{cs}}(\lambda)$ is the β-carotene absorption coefficient in the stratum corneum. The β-carotene absorption coefficient in the stratum corneum is given by

$$\mu_{a_{cs}}(\lambda) = \frac{\varepsilon_{car}(\lambda)}{537}c_{cs}, \tag{7.9}$$

where 537 is the molecular weight of β-carotene (g/mol) and c_{cs} is the β-carotene concentration in the stratum corneum (g/L).

The epidermis total absorption coefficient is given by

$$\mu_{a_2}(\lambda) = \left[\mu_{a_{eu}}(\lambda) + \mu_{a_{ph}}(\lambda)\right]\vartheta_m + \left[\mu_{a_{base}}(\lambda) + \mu_{a_{ce}}(\lambda)\right](1 - \vartheta_m), \tag{7.10}$$

where $\mu_{a_{eu}}(\lambda)$ is the eumelanin absorption coefficient, $\mu_{a_{ph}}(\lambda)$ is the pheomelanin absorption coefficient, $\mu_{a_{ce}}(\lambda)$ is the β-carotene absorption coefficient in the epidermis, and ϑ_m is the volume fraction of the epidermis occupied by melanosomes.

The eumelanin absorption coefficient is given by

$$\mu_{a_{eu}}(\lambda) = \epsilon_{eu}(\lambda)c_{eu}, \tag{7.11}$$

where c_{eu} is the eumelanin concentration (g/L).

Similarly, the pheomelanin absorption coefficient, $\mu_{a_{ph}}(\lambda)$, is computed by multiplying its molar absorption (extinction) coefficient, by its concentration, c_{ph}. Also, the absorption coefficient $\mu_{a_{ce}}$ is obtained by replacing c_{cs} by the concentration of β-carotene in the epidermis, c_{ce}, in Equation 7.9.

The papillary dermis total absorption coefficient is given by

$$\mu_{a_3} = \left[\mu_{a_o}(\lambda) + \mu_{a_d}(\lambda) + \mu_{a_{cd}}(\lambda) + \mu_{a_{bil}}(\lambda)\right]\vartheta_p + \mu_{a_{base}}(\lambda)(1 - \vartheta_p), \quad (7.12)$$

where $\mu_{a_o}(\lambda)$ is the oxyhemoglobin absorption coefficient, $\mu_{a_d}(\lambda)$ is the deoxyhemoglobin absorption coefficient, $\mu_{a_{cd}}(\lambda)$ is the β-carotene absorption coefficient in the dermal layers, $\mu_{a_{bil}}(\lambda)$ is the bilirubin absorption coefficient, and ϑ_p is the volume fraction of the papillary dermis occupied by blood.

The β-carotene absorption coefficient in the dermal layers is obtained by replacing c_{cs} by the concentration of β-carotene in the dermal layers (c_{cd}) in Equation 7.9. Also, recall that the volume fractions of blood vary within the dermis tissue (Section 4.1). Hence, to compute the reticular dermis total absorption coefficient, μ_{a_4}, the volume fraction ϑ_p is replaced by ϑ_r (volume fraction of the reticular dermis occupied by blood) in Equation 7.12. The oxyhemoglobin absorption coefficient is given by

$$\mu_{a_o}(\lambda) = \frac{\varepsilon_o(\lambda)}{66500} c_{hb} * \Lambda, \quad (7.13)$$

where 66500 is the molecular weight of hemoglobin (g/mol), c_{hb} is the concentration of hemoglobin in the blood (g/L), and Λ is the ratio of oxy-hemoglobin to the total hemoglobin concentration.

Similarly, the deoxyhemoglobin absorption coefficient is computed using its molar absorption (extinction) coefficient, and replacing Λ by $(1 - \Lambda)$ in Equation 7.13. Incidentally, the volume fractions ϑ_m, ϑ_p, and ϑ_r are usually provided in terms of percentages, which are converted to values between zero and one before their use in Equations 7.10 and 7.12.

Finally, the bilirubin absorption coefficient is given by

$$\mu_{a_{bil}}(\lambda) = \frac{\varepsilon_{bil}(\lambda)}{585} c_{bil}, \quad (7.14)$$

where 585 is the molecular weight of bilirubin (g/mol) and c_{bil} is the bilirubin concentration (g/L).

7.4 IMPLEMENTATION ISSUES

As described earlier, to perturb the ray reflected on the skin surface, Krishnaswamy and Baranoski [147] used angular displacements obtained from the surface structure proposed by Trowbridge and Reitz [251]. Alternatively, a warping function based on an exponentiated cosine function [22] could be tested. Earlier experiments by Blinn [34] showed that curves resulting from the function proposed by Trowbridge and Reitz and curves resulting from exponentiated cosine functions have a similar shape. Incidentally, the formula for the polar perturbation angle, θ_s, given in the original paper describing the BioSpec model [147] presents a typographical error, namely the position of the exponent $1/2$. This error was fixed in Equation 7.1 presented earlier.

The formula for the polar perturbation angle, θ_s, represents only an approximation [146] because the integration of function proposed by Trowbridge and Reitz and its subsequent inversion to obtain θ_s is not computationally practical. We remark, however, that it is possible to use a data-driven mechanism (such as the table lookup used to simulate the subsurface scattering the stratum corneum and epidermis layers) to compute the perturbation angle. Such an approach would have the benefits of being more accurate as it would use values provided by the original function proposed by Trowbridge and Reitz, and being faster than applying Equation 7.1 on the fly.

In the original implementation of the BioSpec model, two refractive indices were used in a reverse order. As a result, some of the graphs presented in the original publication that described the model [147] were affected by this implementation error. These graphs were regenerated and their correct versions are presented in Figures 7.2 and 7.5.

7.5 STRENGTHS AND LIMITATIONS

BioSpec is the first local lighting model capable of computing both scattering and spectral quantities for skin specimens. Its implementation based on standard Monte Carlo methods enables its straightforward incorporation into image-synthesis pipelines. The algorithmic simulations performed by the model, however, are time consuming and may represent a bottleneck in an image-synthesis pipeline. Alternatively, these simulations can be run offline, and the quantities computed by the model stored and reconstructed during rendering using mathematical techniques such as principal component analysis [101] and regression analysis [221]. These techniques have been successfully used in the efficient reconstruction of spectral data sets for organic [30] and inorganic materials [141].

One of the main apparent challenges for users of comprehensive first principle models such as BioSpec is the relative large number of parameters used by the model. Most of these parameters, however, correspond to biophysical quantities that are not normally subjected to change, and, therefore, can be kept constant. For example, only a small subset of parameters need to have their values modified by the user (Table 7.1) so that different skin-pigmentation levels can be simulated.

Table 7.1 A Summary of the Data Used in the Evaluation of the BioSpec Model. Parameters in Boldface Correspond to User-Specified Data Values

Parameter	Symbol	Default Value	Source
Radius of collagen fibers (nm)	r_f	25	[157]
Refractive index of stratum corneum	η_{sc}	1.55	[64]
Refractive index of epidermis	η_{ep}	1.4	[254]
Refractive index of papillary dermis	η_{pd}	1.36	[128]
Refractive index of reticular dermis	η_{rd}	1.38	[128]
Refractive index of collagen fibers	η_f	1.5	[126]
Thickness of stratum corneum (cm)	h_{sc}	0.001	[11]
Thickness of epidermis (cm)	h_{ep}	0.01	[11]
Thickness of papillary dermis (cm)	h_{pd}	0.02	[11]
Thickness of reticular dermis (cm)	h_{rd}	0.18	[11]
Concentration of melanin in the melanosomes (g/L)	c_{eu}	80	[248]
Concentration of pheomelanin in the melanosomes (g/L)	c_{ph}	12	[248]
Concentration β-carotene in the stratum corneum (g/L)	c_{cs}	2.1^{-4}	[152]
Concentration β-carotene in the epidermis (g/L)	c_{ce}	2.1^{-4}	[152]
Concentration β-carotene in the blood (g/L)	c_{cd}	7.0^{-5}	[152]
Concentration of hemoglobin in the blood (g/L)	c_{hb}	150	[91,160]
Concentration of bilirubin in the blood (g/L)	c_{bil}	0.05	[211]
Epidermis occupied by melanosomes (%)	ϑ_m	5.2	[126]
Papillary dermis occupied by blood (%)	ϑ_p	1.2	[12]
Reticular dermis occupied by blood (%)	ϑ_r	0.91	[12]
Ratio of oxy/deoxy hemoglobin (%)	Λ	75	[12]
Aspect ratio of folds	σ	0.75	[244, 247]

On the scientific side, there is still room for improvement in the BioSpec formulation. For example, similarly to previous models [110, 132, 180], shadowing and masking effects were not taken into account by the BioSpec model. Also, although the simulation of surface reflection performed by the BioSpec model accounts for biological factors and uses a closer approximation to the skin mesostructures than approaches that are based on the use of microfacets [110, 232], its generalization requires a more rigorous mathematical treatment [146]. Furthermore, in the BioSpec formulation, it is assumed that all rays are reflected at the dermis and hypodermis interface. Such an assumption may need to be relaxed as it has not been fully verified through actual experiments.

The BioSpec model was evaluated through quantitative and qualitative comparisons of modeled results with spectral and scattering measured data. Figure 7.2 presents quantitative comparisons of modeled reflectance curves provided by the BioSpec model with actual measured curves provided by Vrhel et al. [270] (available in the North Carolina State University [NCSU] spectra database). According to the subjective description of the specimen provided in the database, Krishnaswamy and Baranoski [147] set the pigmentation levels, represented by the volume fraction (ϑ_m) of the epidermis occupied by melanosomes to 4.1 and 9.5% for a lightly and moderately pigmented specimen, respectively. The measured curves were obtained considering an angle of incidence $\theta_i = 45°$. Vrhel et al. [270] attempted to reduce the specimens specularities by changing their orientation [270], and converted the resulting curves, originally obtained in terms of radiance, to (diffuse) reflectance values. The BioSpec-modeled reflectance curves, however, include both diffuse and

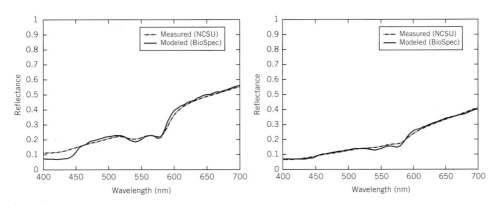

FIGURE 7.2

Comparison of modeled reflectance curves provided by the BioSpec model with actual measured curves available in the NCSU spectra database [270]. Left: lightly pigmented skin specimen (NCSU file 113). Right: moderately pigmented specimen (NCSU file 82).

specular reflectance components, i.e., the target specimens are not assumed to be a perfect Lambertian reflector.

Figure 7.3 presents quantitative comparisons between modeled and actual measured transmittance curves for the stratum corneum and epidermis tissues of two specimens, a moderately and a heavily pigmented one. The measured transmittance data for human skin available in the scientific literature is usually limited to separated skin layers. Krishnaswamy and Baranoski [147] used measured curves (with a reported tolerance of $\approx \pm 5\%$) provided by Everett et al. [82], which were obtained at a normal angle of incidence ($\theta_i = 0°$) [82, 283]. Everett et al. [82] reported thickness values for the moderately pigmented ($h_{sc} = 0.0017$ cm and $h_{ep} = 0.0025$ cm) and the heavily pigmented ($h_{sc} = 0.0023$ cm and $h_{ep} = 0.0021$ cm) specimens. Based on their description of the specimens, Krishnaswamy and Baranoski [147] set ϑ_m to 9.5 and 38% for the lightly and heavily pigmented specimens, respectively.

The quantitative discrepancies between modeled and measured results may be due in part to the fact that the some parameters used in the simulations had to be estimated based on the subjective description of the specimens (e.g., Caucasian or Asian), and other parameters correspond to average values available in the literature (e.g., the refractive indices for the skin layers). Hence, the comparisons between modeled and measured curves provide an upper bound for the accuracy of the simulations; i.e., a more precise quantitative evaluation would also require the measurement of the biophysical characteristics of the specimen to be used as input for the model. Furthermore, the absorption curves used in the simulations were obtained under in-vitro conditions, i.e.,

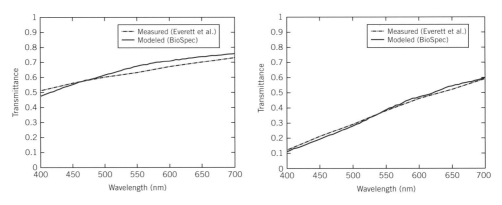

FIGURE 7.3

Comparison of modeled transmittance curves (for the stratum corneum and epidermis tissues) provided by the BioSpec model with actual measured curves provided by Everett et al. [82]. Left: moderately pigmented specimen. Right: heavily pigmented specimen.

they are affected by solvent-related spectral shifts and sieve and detour effects (Section 2.3.3). Although, in theory, these absorption curves could be scaled to account for differences between in-vitro and in-vivo values, in practice such an adjustment is constrained by the scarcity of experimentally obtained scale factors for skin chromophores.

Figure 7.4 shows a comparison between modeled and actual measured skin BRDF data provided by Marschner et al. [166]. Because the BioSpec model provides spectral readings, Krishnaswamy and Baranoski [147] integrated spectral-modeled BRDF values over the visible domain to obtain data that could be compared with the measured BRDF data provided by Marschner et al. [166]. In addition, based on the description of the lightly pigmented specimen provided by Marschner et al. [166], Krishnaswamy and Baranoski [147] set $\vartheta_m = 2.5\%$ in these experiments. As illustrated by the measurements provided by Marschner et al. [166] (Figure 7.4 (top)), the BRDF of skin specimens presents an angular dependence, and it becomes more diffuse for small angles. Figure 7.4 (bottom) shows that the BioSpec model can represent this angular dependency, and the modeled BRDF curves generally agree with the measured BRDF curves provided by Marschner et al. [166]. The most noticeable quantitative discrepancies are observed for the larger angle of incidence, namely $\theta_i = 60°$. It is worth noting, however, that besides the previously mentioned factors that quantitatively affect the modeled curves, one should also consider the sources of noise in the measurements performed by Marschner et al. [166], which include deviations in the specimen's normal estimation and spatial variations in the measured BRDFs.

Because quantitative comparisons may be affected by inherent difficulties to characterize the specimens used in the actual measurements, the evaluation of a model may also be supported by qualitative comparisons. In the case of the BioSpec model, promising evidence of its predictive capabilities

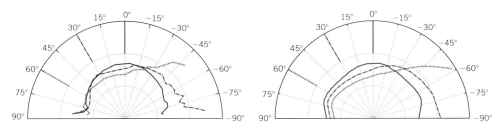

FIGURE 7.4

Comparison of BRDF curves for a lightly pigmented specimen, considering three different angles of incidence: $\theta_i = 0°$, $\theta_i = 30°$, and $\theta_i = 60°$. Left: actual measured BRDF curves provided by Marschner et al. [166]. Right: modeled BRDF curves provided by the BioSpec model.

was provided by the qualitative agreement between modeled results and observed phenomena. For example, the overall reflectance of human skin presents interesting features. As expected, darker skin (characterized by higher volume fractions of epidermis occupied by melanosomes) reflects less light than lighter skin. However, lightly pigmented skin presents a characteristic "W" shape in the reflectance curves between 500 and 600 nm [12]. Oxygenated hemoglobin is responsible for this feature, which can be accentuated as the proportion of oxyhemoglobin with respect to total hemoglobin increases [289]. The graphs presented in Figure 7.5 indicates that the BioSpec model can capture these optical characteristics of human skin. Another example refers to skin BRDF. Skin specimens characterized by thin and numerous folds (e.g., young or hydrated specimens) present a directional behavior stronger than specimens with wider but fewer folds (e.g., old or dry specimens) [166, 244, 247]. The former case corresponds to folds with lower aspect ratio, whereas the later case corresponds to folds with a higher aspect ratio [244, 247]. Figure 7.6 presents modeled BRDF curves for two angles of incidence, namely $\theta_i = 15°$ and $\theta_i = 45°$, obtained by varying the parameter σ associated with the folds' aspect ratio. These curves show that the BioSpec model can qualitatively simulate the variation in the scattering behavior of skin specimens associated with changes in the aspect ratio of the stratum corneum folds.

Besides the experimental evaluation, Krishnaswamy and Baranoski [147] rendered images to illustrate the applicability of the BioSpec model in the

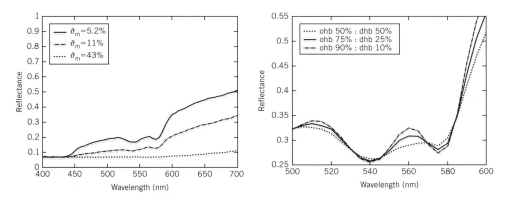

FIGURE 7.5

Comparison of modeled spectral curves provided by the BioSpec model ($\theta_i = 45°$) considering the variation of biological parameters. Left: volume fractions of epidermis occupied by melanosomes (ϑ_m). Right: ratio of oxygenated (ohb) to deoxygenated (dhb) hemoglobin in the dermal layers.

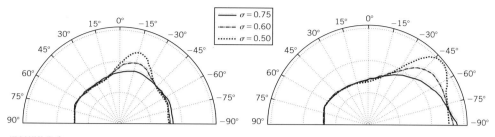

FIGURE 7.6

Comparison of modeled BRDF curves provided by the BioSpec model considering variations on the aspect ratio (σ) of the stratum corneum folds. Left: $\theta_i = 15°$. Right: $\theta_i = 45°$.

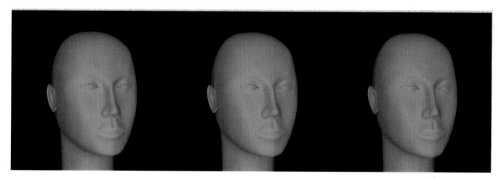

FIGURE 7.7

Images illustrating spectral simulations of jaundice symptoms. The simulations were performed using the BioSpec model [147], and considering the following values for the concentration of bilirubin in the blood: $c_{bil} = 0.05\,\text{g/L}$ (left), $c_{bil} = 0.5\,\text{g/L}$ (center), and $c_{bil} = 3.0\,\text{g/L}$ (right).

spectral simulation of medical conditions, e.g., jaundice (Figure 7.7) and erythema (Figure 7.8), associated with changes in the biophysical parameters. They also rendered images to highlight an aspect for which measured data is scarce, namely the translucency of skin tissues (Figure 7.9). The transmission of light through the whole skin can be observed (in vivo) in body parts with a thin or absent hypodermis, such as ears, eye lids, and fingers. In these areas, the behavior of the transmitted light is near Lambertian to the point where no internal structure can be noticeable [210].

Finally, we remark that BioSpec is a data-driven model. As more data becomes available for the scattering properties of various skin layers or for the spectral properties of additional chromophores, the accuracy of the BioSpec model will increase. However, the predictive simulation of skin appearance

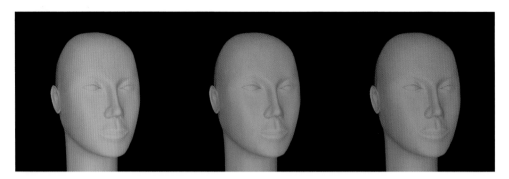

FIGURE 7.8

Images illustrating spectral simulations of erythema conditions. The simulations were performed using the BioSpec model [147], and considering the following values for the volume fractions of the papillary and reticular dermis occupied by blood: $\vartheta_p = 1.2\%$ and $\vartheta_r = 0.91\%$ (left), $\vartheta_p = 2.7\%$ and $\vartheta_r = 0.3\%$ (center), and $\vartheta_p = 3.6\%$ and $\vartheta_r = 0.4\%$ (right).

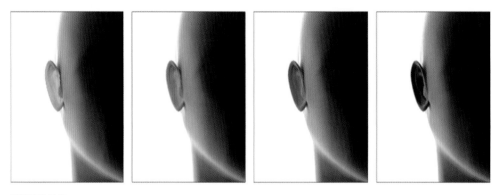

FIGURE 7.9

Images generated using the BioSpec model [147] to show variations in the translucency of skin tissues associated with different levels of melanin pigmentation, represented by the volume fraction (ϑ_m) of epidermis occupied by the melanosomes. From left to right: $\vartheta_m = 1.9\%$, $\vartheta_m = 5.2\%$, $\vartheta_m = 12\%$, and $\vartheta_m = 42\%$.

changes as a result of time-dependent biophysical phenomena (e.g., tanning) will likely require a lower level of abstraction in which key tissue layers, such as the epidermis, would have to be subdivided into its sublayers. Clearly, such a modeling refinement will be also bound by data availability.

Diffusion approximation approach

The diffusion theory [123] has been used in different fields to solve problems requiring an approximate solution for the equation describing light transport in highly scattering media. In tissue optics, it has provided the foundation for several light interaction models aimed at medical applications (Chapter 5). Although the accuracy of the results provided by the diffusion approximation are bound by the optical characteristics of the target material, its amenability to analytic manipulation and relative low computational costs have also motivated its use in computer-intensive applications such as the realistic modeling of material appearance.

In this chapter, we examine a class of local lighting models whose design was based on the application of the diffusion theory. These models were developed primarily to be incorporated into image-synthesis pipelines. Our review starts with the model proposed by Jensen et al. [132], henceforth referred to as DT model. This model applies the general concept of the bidirectional scattering-surface reflectance distribution function (BSSRDF) (Section 2.6.2) to describe the propagation of light from one point on a surface to another. The performance of the DT model was later improved by Jensen et al. [130] through the incorporation of a two-pass hierarchical algorithm. As these works represent the first appearance modeling efforts to use the diffusion theory within the computer graphics field, they are the focal point of the discussions presented in this chapter.

However, the application of the diffusion theory in the modeling of material appearance has evolved beyond these models. Accordingly, the discussion of models based on the diffusion-approximation approach proceeds to address relevant works that built on and extended the techniques used in the DT model. Most of the models examined in this chapter have been applied to the rendering of a variety of materials, from marble to milk. In this book, they are examined with respect to theoretical and practical issues involving their use in the generation of realistic images depicting human skin.

8.1 OVERVIEW

Jensen et al. [132] observed that because of the effects of repeated multiple scattering, the light distribution tends to become symmetric (equal in all directions) and blurred in highly scattering media. Because the diffusion theory does not have a general analytical solution for finite media, they modeled subsurface reflection as a semiinfinite medium. They used the diffusion approximation for isotropic media known as the dipole method (Section 5.3). It consists of two point sources placed relative to the surface (Figure 8.1), one representing the positive real light located below the surface, and the other representing a negative virtual light positioned above the surface. Using this approximation, they compute the radiant exitance at the propagation point, x_p, separated by a certain distance, d_p, from the incidence point, x_i.

The DT model has four input parameters: the absorption coefficient, the reduced scattering coefficient, the diffuse reflectance, and the index of refraction. In order to determine the values of these parameters for various materials, they used a 3-CCD (charge coupled device) video camera to observe the radiant exitance across the surface of the material. They then used the diffusion theory to compute the absorption coefficient and reduced scattering coefficient. In their follow-up work, Jensen and Buhler [130] reduced the space of parameters of the DT model to the diffuse reflectance and an average scattering distance. In addition, this follow-up work presented a more efficient

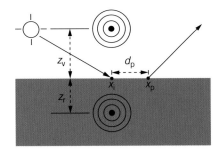

FIGURE 8.1

Sketch describing the geometry used in the transformation of an incoming ray into a dipole source according to the diffusion approximation. The distances from the real and virtual dipole sources to the air and material interface are given by z_r and z_v, respectively, and the distance from the incidence point x_i to the propagation point x_p is giving by d_p.

scheme for sampling the incident flux as a result of subsurface scattering without altering the fundamental concepts used in the original work.

8.2 SCATTERING SIMULATION

The DT model is primarily a subsurface scattering model, and surface reflection is not directly addressed in its formulation. However, it can be combined with other models in order to handle the surface reflection. For example, Jensen and Buhler [131] used a Lambertian model in conjunction with the model proposed by Torrance and Sparrow model [249] in order to improve the realistic appearance of a human face rendered using the DT model.

The DT model is formally defined by Jensen et al. [132] as a sum of a single-scattering term and the diffusion approximation. Jensen et al. proposed the use of an analytical approximation to account for single-scattering events, which may occur when a refracted incoming ray and an outgoing ray intersect (Figure 8.2). It consists in computing the light transport integral over a selected portion of the path followed by light along the direction of propagation using a formulation based on the analytical solution for first-order scattering proposed by Hanrahan and Krueger (Section 6.1.2).

For the diffusion approximation, Jensen et al. proposed an algorithmic approach which can be summarized as follows:

1. generate a set of sampling points on the surface of the model,

2. for each sampling point, compute the incident irradiance using any lighting algorithm, and

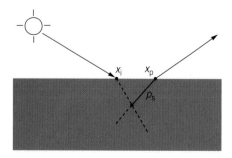

FIGURE 8.2

Sketch describing the geometry used to account for single scattering, which is computed as an integral over the path p_s along the direction of propagation. The incidence and propagation points are represented by x_i and x_p, respectively.

3. for each sample point, compute the radiant exitance using the following equation which takes into account the geometry presented in Figure 8.1

$$dM_o(x_o) = d\Phi_i(x_i)\frac{\gamma'}{4\pi}\left(C_1\frac{e^{-\mu_{tr}d_r}}{d_r^3} + C_2\frac{e^{-\mu_{tr}d_v}}{d_v^3}\right), \tag{8.1}$$

with C_1 expressed as

$$C_1 = z_r\left(\mu_{tr} + \frac{1}{d_r}\right)$$

and C_2 expressed as

$$C_2 = z_v\left(\mu_{tr} + \frac{1}{d_v}\right),$$

where $d\Phi_i(x_i)$ is the incident flux at sampling point, γ' is the reduced albedo, μ_{tr} is the effective transport extinction coefficient, d_r is the distance from x_p to the real dipole light source, d_v is the distance from x_p to the virtual dipole light source, z_r is the distance from x_i to the real dipole light source, and z_v is the distance from x_i to the virtual dipole light source.

The reduced albedo, γ', is given by

$$\gamma' = \frac{\mu_s'}{\mu'}, \tag{8.2}$$

where μ_s' corresponds to the reduced scattering coefficient (Section 5.3), and the reduced attenuation coefficient, μ', corresponds to

$$\mu' = \mu_s' + \mu_a. \tag{8.3}$$

Incidentally, in the original article describing the DT model [132], this quantity is referred as the reduced extinction coefficient.

The effective transport extinction coefficient, μ_{tr}, can be expressed as

$$\mu_{tr} = \sqrt{3\mu_a\mu'}. \tag{8.4}$$

The distance to the real light source, d_r, is given by

$$d_r = \sqrt{d_p^2 + z_r^2}, \tag{8.5}$$

where d_p is the distance from x_p to x_i.

Similarly, the distance to the virtual light source, d_v, is given by

$$d_v = \sqrt{d_p^2 + z_v^2}. \tag{8.6}$$

Finally, the distance from the virtual dipole light to the surface, z_v, is given by

$$z_r = \frac{1}{\gamma'}, \tag{8.7}$$

and the distance from the real dipole light to the surface, z_r, can be expressed as

$$z_v = z_r + 4 \frac{1 + F_{dr}}{1 - F_{dr}} \frac{1}{3\mu'}, \tag{8.8}$$

where the Fresnel term F_{dr} corresponds to the average diffuse Fresnel reflectance. Jensen et al. [132] computed this term using an approximation, proposed by Egan and Hilgeman [78], for the internal diffuse reflectance, which is given by

$$F_{dr} = -\frac{0.5601}{\eta^2} + \frac{0.710}{\eta} + 0.668 + 0.0636\eta_r, \tag{8.9}$$

where η is the refractive index of the medium.

The expression for F_{dr} given in Equation 8.9 corresponds to a curve fit to tabular data provided by Orchard [189].

8.3 IMPLEMENTATION ISSUES

The three main tasks involved in the implementation of the DT model, according to a more compact description provided by Jensen and Buhler [130], are

1. the implementation of the diffusion approximation equations described earlier,

2. the computation of the sample points, and

3. the optimization of the algorithm through the use of an hierarchical technique.

The optimization proposed by Jensen and Buhler [130] consists in use of an hierarchical data structure so that sample points can be selected by their relevance, and points that are distant can be grouped together. More specifically, Jensen and Buhler suggested the use of an octree [215] for the implementation of this hierarchical acceleration mechanism.

The approximation for F_{dr} proposed by Egan and Hilgeman [78], which appears in the articles describing the DT model [132] and its more efficient version [130], has also been applied in the work by Groenhuis et al. [105] in the investigation of scattering and absorption in turbid materials. In both cases, i.e., the papers describing the DT model and the work by Groenhuis et al. cited in these papers, the formula for this approximation presents an error. More specifically, it has the constant -1.44 multiplying the η_r^{-2} term instead of the constant -0.56 that appears in Equation 8.9. Incidentally, the formula given in Equation 8.9 was obtained directly from the expression provided by Egan and Hilgeman [78]. Furthermore, the term η_r, defined as "the relative index of refraction of the medium with the reflected ray to the other medium" in the original description provided by Jensen et al. [132], is defined as "the index of the refraction of the medium" in the works of Egan and Hilgeman [78] and Groenhuis et al. [105].

8.4 STRENGTHS AND LIMITATIONS

The DT model, differently from many works in tissue optics, does not assume that the distance from the point of incidence to the point of propagation is negligible. From a theoretical point of view, this is a valid contribution because such an assumption fails to represent the real behavior of diffusive or translucent materials. In practice, however, the effects of this assumption on the appearance of the materials may not be as significant as the effects resulting from other assumptions such as the homogeneity of the materials. Furthermore, the errors introduced in a scattering simulation by the elimination of the positional argument of the BSSRDF may be mitigated when one applies stochastic simulation approaches. Even though the inclusion of such positional argument may contribute to increase the accuracy of the scattering simulations, such an improvement may be obscured by the use of other approximations, such as the diffusion approximation itself, and other simplifying techniques such as the application of a curve fit formula to compute the Fresnel terms.

The DT model is relatively simple to implement, and it is not as computationally expensive as Monte Carlo-based models. In addition, it can be used to render visually pleasing images. However, similarly to the H–K model, it presents some limitations associated with its generality. It does not take into account properties specific to organic materials. Furthermore, because spectral properties, such as the diffuse reflectance, are actually input parameters to the simulations, the DT model shall be classified only as a scattering model.

The diffusion approximation used in the DT model requires the computation of the reduced scattering coefficient (Section 5.3), which, in turn, uses as a parameter the asymmetry factor, g, of the phase function. Jensen et al. [132] applied the HGPF and used $g = 0.85$ for the whole skin. In the second article, Jensen and Buhler [130] used the HGPF with the asymmetry values provided by van Gemert et al. [264]. Besides the direct effect on the predictability and accuracy of the scattering simulations (Section 5.5), the use of the HGPF and the selection of asymmetry factors may have further theoretical implications. For instance, Jensen and Buhler [130] state that the diffusion approximation has been shown by Furutso [98] to be accurate when

$$\frac{\mu_a}{\mu_a + \mu_s} << 1 - g^2.$$

In order to apply this relationship, one needs to know the values of its terms, which in turn come from measured data. As discussed earlier (Chapter 5), the values of g available in the tissues optics literature, including the values mentioned earlier, usually come from fitting approaches, i.e., they are already approximations.

The DT model considers the entire skin structure as one medium. As described in Chapter 4, skin is heterogeneous and multilayered, with each of the layers having different biological and optical properties. These properties have a direct impact on the effectiveness of the diffusion theory. For example, the diffusion approximation is not suitable when the scattering is mostly in the forward direction [92, 98, 285]. As highlighted in Section 4.3, the measurements performed by Bruls and van der Leun [38] demonstrated that both the stratum corneum and the epidermis tissues are highly forward scattering media. Furthermore, the diffusion theory is not applicable when the absorption coefficient is not significantly smaller than the scattering coefficient for turbid media. Recall that human skin is characterized by the presence of pigments, such as melanin particles, which have a significant absorption cross section [42].

The evaluation of the DT model and its variant was based solely on visual inspection. For example, Jensen et al. [132] compared images generated using a bidirectional reflectance distribution function (BRDF) model (not specified by Jensen et al. [132]) with images generated using the DT model. Because

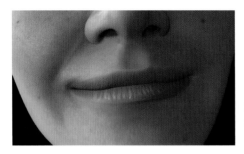

FIGURE 8.3

Images of a textured face before (left) and after (right) the addition of the subsurface scattering simulation provided by the DT model [130].

comparisons with actual BSSRDF data of any organic (or inorganic) material were not provided, neither the accuracy nor the predictability of the DT models can be verified. Despite these issues, the DT model has been successfully used in the generation of believable images of human beings used in entertainment applications (Figure 8.3). Moreover, the development of the DT model has addressed the positional assumption made by the other models, and it has raised relevant theoretical issues concerning subsurface scattering simulations.

8.5 EVOLUTION OF DIFFUSION APPROXIMATION–BASED MODELS

The limitations mentioned in the earlier section have been acknowledged by Donner and Jensen [68] in a subsequent publication describing a hybrid model for simulating the scattering profile of translucent materials. This models uses the multipole method, which uses multiple dipoles, and a variation of the K-M theory. The multipole method, originally proposed in biomedical literature [54, 272], accounts for the refractive index mismatches at the material boundaries. Donner and Jensen [68] also combined the effects of reflectance and transmittance from each layer of a material (given by the multipole method) using the K-M theory. Rather than convolve the reflectances and transmittances, they performed their calculation in frequency domain, where the convolutions become products and can be further simplified as they form a geometric series. This model was evaluated using limited set of comparisons with results provided by a Monte Carlo simulation (for which no formulation details were provided by Donner and Jensen), and used to generate a compelling renderings of human skin.

Donner and Jensen in a continuing work [69] applied the multipole method in conjunction with a spectral model to specifically render the appearance of human skin. They considered skin to be formed by two layers, the epidermis and dermis, and used simple functions to approximate the absorption coefficients of hemoglobin and melanins. These were, in turn, used to generate the spectral absorption and scattering coefficients for each of the two layers which are then convolved using the multipole method. Although their model contains several biophysically based parameters, such as the thicknesses of the skin layers, they reduced the user-set parameters to only three, namely hemoglobin fraction (the percentage of hemoglobin in the dermis), melanin fraction (the percentage of melanin in the epidermis), and the melanin type blend (the ratio of pheomelanin to eumelanin). Incidentally, a similar approach was previously used by Krishnaswamy and Baranoski [147] in their description of the BioSpec model; i.e., a group of characterization parameters that are specimen independent, such as refractive indices, were kept fixed. In their case, however, the default values for all parameters, fixed and user-set, were provided to facilitate the reproduction of their modeled results.

This modeling framework proposed by Donner and Jensen [69] has two key differences with respect to previous DT models described in this chapter. First, the input parameters mentioned earlier are biologically meaningful. This facilitates the qualitative evaluation of the predictive capabilities of the model. Second, the model was also quantitatively evaluated by comparing modeled reflectance curves with measured curves provided by Vrhel et al. [270] (Figure 8.4). Recall, however, that only a subjective description is provided for the individuals considered in the measurements, which makes the choice of parameter values to be used in the evaluation of a model subject to user input. Although the relatively small number of user-specified parameters used in their model facilitates the selection of values that provide the best match between modeled and measured curves, their parameter set is not detailed enough to provide an appropriate characterization of different individuals with the same subjective description (e.g., Caucasian or African-American). This also depends on other biophysical parameters such as the thickness of the tissues and the roughness of the skin surface. Hence, the results of their comparisons should be seen as a loose upper bound for the fidelity of their model, especially considering that the absorption data used in their simulations was derived from formulas used to approximate measured absorption spectra. It is also important to note that a key measurement of appearance component was not evaluated by Donner and Jensen [69]. More specifically, they have not compared modeled scattering profiles to measured ones. Although one has to acknowledge the relative scarcity of reliable skin-scattering data, we remark

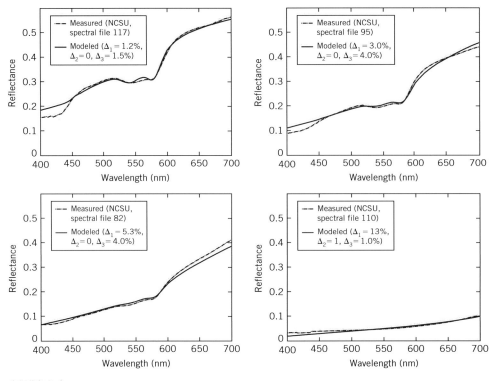

FIGURE 8.4

Comparison of modeled skin diffuse reflectance curves obtained using Donner and Jensen's approach [69] with measured curves (available at the NCSU database [270]). Modeled curves were obtained varying the following parameters: hemoglobin fraction (Δ_1), melanin fraction (Δ_2) and melanin type blend (Δ_3). Top left: Caucasian individual. Top right: dark Caucasian individual. Bottom left: Asian individual. Bottom right: African-American individual.

that other modeled scattering profiles (Section 7.5) have been previously compared with measured skin BRDF data [164–166].

Donner and Jensen [70] also proposed an extension which combined photon tracing [226] with a diffusion model to simulate a variety of effects because of the presence of internal occluding geometry in translucent materials. The algorithm consists of two passes. In the first pass, photons (rays) are traced from light sources, and when a photon hits a translucent material, it is refracted and propagated in the volume of that material until

1. the photon hits an exiting interface, in which case it is propagated,

2. the photon is absorbed by the medium or by some occluding geometry embedded in the medium, or

3. the photon is scattered, in which case the photon position, power, and plane of incidence are stored. The photon is then propagated and continues to be scattered until events 1 or 2 occur, and it is only stored if it exits the medium and reenters it.

During rendering, in order to estimate the radiant exitance at a point on the surface of a translucent object, the radiance contribution from all stored photons is considered. Each stored photon is treated as a diffusion source, which approximates multiple scattering. Single scattering is computed by tracing rays in the volume and evaluating the radiance from the stored photons.

The dipole and multipole diffusion approximation assume that the material surface is flat and infinite in extent. Donner and Jensen [70] realized that many 3D objects are represented by polygonal meshes that rarely satisfy this assumption. Therefore, they introduced the concept of quadpole diffusion approximation made up of two real and two virtual sources. In addition, in order to reduce the computational time required for the evaluation of contributions provided by millions of sources, they used a hierarchical data structure similar to the one used by Jensen and Buhler [130]. After completing photon tracing, the photons are collected into the voxels of an octree with each voxel approximately representing the photons in that volume. Each voxel stores the total power, average position, and average plane of incidence. This voxel approximation is constructed hierarchically. During rendering, the octree is traversed until either a node within a certain threshold of solid angle (as viewed from the point of shading) or a leaf node is reached. The representative voxel data is used for the diffusion computation. The validation of their work was also performed through the visual inspection of several sets of images.

Subsequently, Donner et al. [71] attempted to address one of the limitations of previous diffusion theory–based models, namely the assumption that skin layers are homogenous. More specifically, they used 2D parameter maps to account for biophysical variations within these tissues. Donner et al. presented a spectral skin-shading model, which expanded on the work by Donner and Jensen [73] and D'Eon et al. [61]. Their model consists of two main layers (epidermis and dermis) along with an infinitesimally thin absorbing layer between them. The following parameters are considered for the epidermis: melanin fraction, eumelanin and pheomelanin ratio (or melanin type blend), hemoglobin fraction, and β-carotene fraction (the percentage of β-carotene in the epidermis). The only parameter used for the dermis is the hemoglobin fraction. They also used a parameter, called oiliness, to scale the surface reflectance calculation made using the Torrance Sparrow model [249].

Donner et al. [71] remarked that the interface between the epidermis and the dermis is corrugated. Because the interface between these layers in their

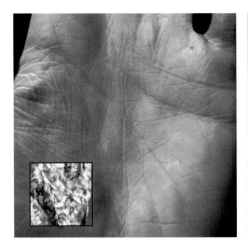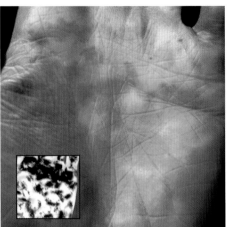

FIGURE 8.5

Images simulating appearance changes resulting from blood flow (hemoglobin concentration) variations after clenching and releasing a hand. The parameter maps (scaled 20×) for hemoglobin employed in the simulation framework proposed by Donner et al. [71] are shown in the insets. Donner, C., Weyrich, T., D'eon, E., Ramamoorthi, R., and Rusinkiewicz, S. "A layered, heterogeneous reflectance model for acquiring and rendering human skin" *ACM Transactions on Graphics* 27, 5 (2008), 1–12. © 2010 Association for Computing Machinery, Inc. Reprinted by permission.

model is smooth, they attempted to account for the roughness between the layers by including some amount of hemoglobin in the epidermis. They also use a variable to control the ratio of oxyhemoglobin to deoxyhemoglobin. It is worth noting, however, that the value for this variable was fixed at 0.7, along with the refractive indices for the epidermis (1.4) and dermis (1.38), and the thickness of the epidermis (0.25 mm). The dermis was assumed to be semiinfinite for thick surfaces. The absorption in the infinitesimally thin layer between the epidermis and dermis was assumed to come from melanin, and it was fixed at 17.5% of the total absorption of melanin in the epidermis. Donner et al. [71] generated several sets of images to illustrate the effect of their 2D parameter maps on rendered output as seen in Figure 8.5. They also presented a method for creating these maps using a customized measurement apparatus.

Simulation challenges

9

As it has been illustrated throughout this book, light transport algorithms form the core of simulation frameworks aimed at the realistic rendering of human skin. The development of such algorithms, however, relies on the application of a scientifically sound methodology that involves data gathering, modeling, and evaluation procedures. Although this development process seems straightforward, it is affected by a number of theoretical and practical constraints. In order to move forward and increase the efficacy of existing simulation frameworks, it is necessary to find alternatives to effectively address these constraints. However, this is not a trivial task. In several instances, the root of the problem is outside of the computer science domain, and its solution has been eluding scientists for decades.

Some of these constraints have been mentioned previously in this book. In this chapter, we bring them to the center stage. More specifically, we provide an overview of theoretical and practical issues related to the development and deployment of algorithms for the simulation of skin appearance. We remark that light and skin interactions are object of extensive study in a myriad of fields. Accordingly, the development of more robust simulation frameworks is likely to have a positive impact in these fields. Hence, there are tangible opportunities for collaborations among different research communities. Such collaborations may be instrumental to overcome the simulation challenges in this area.

9.1 INPUT DATA ISSUES

As described in the previous chapters, there is a number of local lighting models for human skin that can be classified as biophysically based. One of the major problems involving the use of these models refers to the availability of accurate input data. An example of data scarcity relates to the refractive indices of organic materials. Ideally, one should use wavelength-dependent quantities. In practice, most refractive indices available in the literature correspond to either values measured at a specific wavelength or values averaged over a wide spectral region. These approximations may introduce significant errors in the simulations as the refractive indices of many natural materials vary considerably. Another example relates to the absorption spectra of various skin pigments. These curves correspond to in-vitro values, i.e., they are not measured in the tissues, but dissolved in a solvent. Consequently, the in-vitro values may present spectral shifts with respect to in-vivo values because of the influence of the solvent. These values may be also affected by the limited sensitivity of experimental measurement procedures [81]. Clearly, more research efforts need to be directed toward the reliable measurement of fundamental biophysical data so that the predictability of skin optics simulations can be improved. Such efforts will likely include the use of alternative measurement approaches. For example, photoacoustic absorption spectroscopy (PAS) [212] may be used used to mitigate some of the sensitivity limitations of traditional absorption spectroscopy techniques [81].

PAS uses pulsed light to illuminate a sample in an enclosed gas-filled cell. The incident light is absorbed, which causes the sample to enter into an excited state. Deexcitation of the sample can take place in a number of different ways including the reradiation of the absorbed energy as heat. When the sample radiates heat, it also raises the temperature of the surrounding gas and, correspondingly, the pressure inside of the cell. The pulsed nature of the incident light makes the pressure inside the cell to change in a similar manner. The pressure changes result in waves that can be picked up with a microphone. Plotting the relative signal strength generated by the sample at different wavelengths of incident light produces a photoacoustic absorption spectrum, which qualitatively resembles that of an absorption spectrum. Besides, its higher sensitivity, an additional advantage of PAS over other light-absorption spectroscopy, is the ability to obtain the optical and thermal properties of highly scattering solid and semisolid materials such as powders, gels, suspensions, and tissues. We remark, however, that during the deexcitation process, some of the absorbed energy may be channeled into biophysical processes and radiation of nonthermal energy. These processes should be taken into account when converting photoacoustic signals of individual pigments to specific absorption coefficients [81].

9.2 MODELING ISSUES

The current models of light interaction with human skin do not account for the anisotropy of human skin, and usually the simulation of shadowing and masking effects lacks a sound biophysical basis. Although physically based algorithms for the modeling of these phenomena with respect to inorganic materials have been proposed by computer graphics researchers [95, 200], the generalization of these algorithms to organic materials is not straightforward because the simulation of the underlying biophysical processes involves the whole hierarchy of local illumination, i.e., issues at the microscopic (bidirectional reflectance distribution function (BRDF)) mesoscopic (normal distribution) and macroscopic (geometry) levels need to be addressed. In order to obtain predictable solutions for these questions, it will be essential to combine fundamental research efforts on the measurement of surface microrelief data [155, 287] with the careful modeling of skin geometrical details [247].

Accuracy and computational time are often conflicting issues in biophysically based rendering. Although the main goal in this area is the design of accurate and efficient models, sometimes it is difficult to obtain this perfect combination. In order to achieve a higher level of accuracy, it may be necessary to add complexity to a model, which, in turn, may negatively affect its computational performance. However, this is not always the case. For example, phase functions, such as the Henyey–Greenstein phase function (HGPF) and its variations, were originally used in tissue subsurface scattering simulations to fit data measured at specific wavelengths (Chapter 5). Since then, their application has been extended to different organic materials despite the lack of supporting measured data and the fact that their parameters have no biological meaning. It has been demonstrated (Chapter 5) that the use of a data-oriented approach instead of the HGPF increases the accuracy and time ratio of algorithmic subsurface scattering simulations, and contributes to their improved predictability as the simulations are not controlled by ad-hoc parameters.

9.3 EVALUATION ISSUES

In the computer graphics field, a local lighting model is primarily designed for picture-making purposes, and its evaluation is often based on the visual inspection of the resulting synthetic images. Surely, this evaluation approach is affected by factors not directly related to the model. For example, the use of a high-resolution polygonal mesh and the application of albedo and texture maps may substantially contribute to the realistic appearance of rendered skin (Figure 9.1). These aspects, however, correspond to different stages of the image-synthesis pipeline (Chapter 3), and should be evaluated independently.

Alternatively, light transport models are often evaluated through comparison of their results with the results provided by existing models, usually Monte Carlo-based models. However, the details about the design and accuracy of the model used as reference are often not disclosed in technical publications. This reduces the reliability of the comparisons. After all, even if there is a close agreement between the results provided by the new model and the reference model, there is no guarantee that the predictions made by both models are not far from the "real thing."

A more reliable evaluation approach involves direct comparisons between model readings and actual measured data. For example, the BioSpec model (Chapter 7) was tested as a separated unit of the rendering pipeline and the

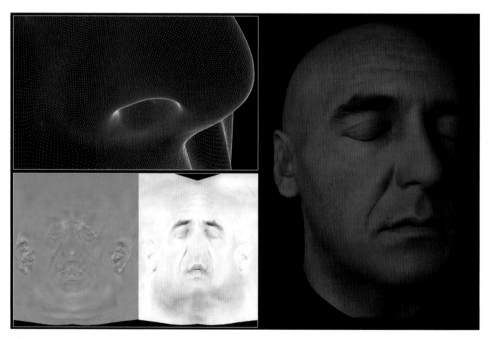

FIGURE 9.1

Sequence of images illustrating a process of creating a human face. Top left: 3D mesh obtained using 3D scanning (close-up of nostril). Bottom row, from left to right: normal map used to give rendered surface subtle detail, and albedo map used to approximate the presence of absorbers (such as hair) on top of the skin. Right: final result rendered using the BioSpec model (Chapter 7) and a spectral path tracer. (Polygonal mesh and textures courtesy of XYZ RGB Inc.)

results were compared with the actual measured data. These comparisons were performed using a virtual spectrophotometer and a virtual goniophotometer (Chapter 3), and reproducing the actual measurement conditions as faithfully as possible. Once a model has been quantitatively evaluated using this approach, its use as a reference for the evaluation of the relative accuracy and performance of other models becomes more reliable because one then knows how far its predictions are from the "real thing."

The direct-evaluation approach, however, is also bounded by the availability of measured data. As mentioned earlier, skin reflectance and transmittance curves can be found in the biomedical literature, but they are usually restricted to a narrow range of measurement conditions. Furthermore, high fidelity comparisons of measured and modeled quantities require both qualitative and quantitative characterization data for the specimen at hand. Usually, only qualitative characterization data is available for skin specimens used in tissue-optics experiments. Although a substantial amount of research has been devoted to the in-vivo characterization of skin specimens, i.e., the quantification of parameters such as pigment content [71, 80, 253, 280] and skin microrelief [155, 195], we remark that the most of this research is based on the use of imaging techniques in conjunction with inversion procedures. Hence, it is important to consider that the quality of the images can be affected by several factors (e.g., the presence of hair, air bubbles, light reflection, and over- and underexposure [106]), and the reliability of inversion procedures depends on models whose evaluation may be affected by the same problems mentioned previously. An alternative for the characterization of specimens under in-vivo conditions may be the use of PAS procedures [17, 269].

In terms of goniometric (BRDF and bidirectional transmittance distribution function [BTDF]) curves, despite efforts of computer graphics researchers [58, 164–166], there is still a shortage of data, and the available data sets present the same limitations outlined for the reflectance and transmittance data sets. The lack of measured data is even more serious with respect to light propagation below the surface. To date, one of the few available data sets corresponds to the measurements performed by Bruls and van der Leun [38], which are limited to two wavelengths in the visible range. Therefore, substantial efforts should be also directed toward the reliable measurement of multispectral surface and subsurface light-propagation profiles.

9.4 PERFORMANCE ISSUES

Performance is the key for many interactive computer graphics applications. For this reason, computationally inexpensive models are routinely chosen for this task over more expensive biophysically based models. In fact, the computational cost has became the Achilles' heel of these models. We remark, however, that many alternatives exist to effectively address this problem. For example, these models, especially those based on stochastic simulations, are highly amenable to parallelization and significant speedup gains can be obtained using a divide and conquer parallel strategy [148]. Another alternative involves the use of robust mathematical methods, such as principal component analysis [30, 121, 243] and regression analysis [120, 141], to obtain compact off-line representations of modeled results. These representations can then be efficiently incorporated into the rendering pipeline during running time, dramatically reducing the computational time overhead of these models.

Finally, we believe that highly parallel graphics hardware and advanced programming application programming interfaces (APIs) may further increase the viability of biophysically based algorithms running at interactive rates. For example, one of the first interactive algorithms for rendering human skin was proposed by d'Eon et al. [61]. They used a technique based on sums of Gaussians to approximate dipole and multipole diffusion (Chapter 8). They were able to implement their algorithm to run on a graphics processing unit (GPU) and generate believable images of human skin at real-time speeds.

Beyond computer graphics applications

10

One of the main driving forces of computer graphics has been the generation of believable images. In the case of human skin, such images can simply result from biophysically inspired simulations, i.e., they can be produced without resorting to sophisticated biophysically based models. For example, the image of a human head depicted in Figure 10.1 was rendered using texture maps and a standard Lambertian model.

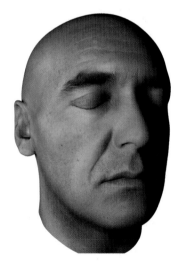

FIGURE 10.1

A human head rendered using a high-resolution geometrical representation and texture maps. The spatial distribution of light is approximated using a Lambertian model, and considering the head lit by an area light source. (Polygonal mesh and textures courtesy of XYZ RGB Inc.).

DOI: 10.1016/B978-0-12-375093-8.00010-1

One may then wonder what practical benefits can arise from investing more time and resources to design and implement biophysically based predictive simulations. After all, for many applications, notably, in the entertainment industry, predictability is not essential. We remark that predictability can make the image-generation process more automatic as it can be controlled by biophysically meaningful parameters. More importantly, predictive simulations have a wider range of scientific and medical applications. Once their fidelity is evaluated through a sound methodology, these simulations can be used not only to generate believable images, but also to provide an in-silico experimental framework for photobiological phenomena that cannot be studied through traditional "wet" measurement procedures. In this way, predictable models can contribute to accelerate the hypothesis generation and validation cycles of research in different scientific domains.

Recall that computer simulations can also assist the visual diagnosis of different skin conditions and diseases, including skin tumors, and the indirect (noninvasive) measurement of the optical properties of skin tissues. These applications involve the noniterative and the iterative use of a theoretical model of light interaction with these tissues. The noniterative approach consists in inverting the model (Section 5.1). In an iterative approach, the optical properties are implicitly related to measured quantities (e.g., reflectance and transmittance). The model is then run iteratively using different input parameters until these quantities are matched. Clearly, the predictability of these simulations is tied to the accuracy of the model being used. This aspect emphasizes the importance of evaluating the models both qualitatively and quantitatively. However, such a thorough evaluation approach is rarely used. Another relevant aspect highlighted in this book is that no single model is superior in all aspects involved in the simulation of light interaction with human skin, and there are still many avenues of research to be explored in this area. These are not limited to substantial improvements on existing frameworks, but also include the solution of open problems such as the predictive simulation of appearance changes triggered by radiation outside the visible range.

From a computer graphics perspective, it is clear that any effort to advance the state-of-the-art of skin appearance rendering will likely depend on the availability of reliable biological data. However, future developments in this area will likely find theoretical and practical applications beyond the image synthesis boundaries. Viewed in this context, we hope that the concise multidisciplinary background on the simulation of light and skin interactions provided in this book may contribute to the establishment of new collaborations among different scientific communities that share similar interests in this important and enduring topic.

References

[1] P. Agache, Assessment of erythema and pallor, in: P. Agache, P. Humbert (Eds.), Measuring the Skin, Springer-Verlag, Berlin, 2004, pp. 591-601.

[2] P. Agache, Main skin physical constants, in: P. Agache, P. Humbert (Eds.), Measuring the Skin, Springer-Verlag, Berlin, 2004, pp. 747-757.

[3] P. Agache, Metrology of the stratum corneum, in: P. Agache, P. Humbert (Eds.), Measuring the Skin, Springer-Verlag, Berlin, 2004, pp. 101-111.

[4] P. Agache, Pigmentation assessment, in: P. Agache, P. Humbert (Eds.), Measuring the Skin, Springer-Verlag, Berlin, 2004, pp. 506-510.

[5] P. Agache, Thermometry and thermography, in: P. Agache, P. Humbert (Eds.), Measuring the Skin, Springer-Verlag, Berlin, 2004, pp. 354-362.

[6] P. Agache, S. Diridollou, Subcutis metrology, in: P. Agache, P. Humbert (Eds.), Measuring the Skin, Springer-Verlag, Berlin, 2004, pp. 410-424.

[7] S. Alaluf, D. Atkins, K. Barret, M. Blount, N. Carter, A. Heath, Ethnic variation in melanin content and composition in photoexposed and photoprotected human skin, Pigment Cell Res. 15 (2002) 112-118.

[8] S. Alaluf, A. Heath, N. Carter, D. Atkins, H. Mahalingam, K. Barrett, R. Kolb, N. Smit, Variation in melanin content and composition in type V and type VI photoexposed and photoprotected human skin: the dominant role of DHI, Pigment Cell Res. 14 (2001) 337-347.

[9] S. Alaluf, U. Heinrich, W. Stahl, H. Tronnier, S. Wiseman, Dietary carotenoids contribute to normal human skin color and UV photosensitivity, J. Nutr. 132 (2002) 399-403.

[10] W. Allen, H. Gausman, A. Richardson, J. Thomas, Interaction of isotropic light with a compact plant leaf, J. Opt. Soc. Am. 59 (10) (1969) 1376-1379.

[11] R. Anderson, J. Parrish, The optics of human skin, J. Invest. Dermatol. 77 (1) (1981) 13-19.

[12] E. Angelopoulou, Understanding the color of human skin, in: B.E. Rogowitz and T.N. Pappas (Eds.), Human Vision and Electronic Imaging VI, SPIE, vol. 4299, Bellingham, Washington, 2001, pp. 243-251.

[13] ANSI, Nomenclature and definitions for illuminating engineering, in: ANSI/IES RP-6-1986, Illuminating Engineering Society of North America, New York, 1986.

[14] M. Arnfield, R. Mathew, J. Tulip, M. Mcphee, Analysis of tissue optical coefficients using an approximate equation valid for comparable absorption and scattering, Phys. Med. Biol. 37 (6) (1992) 1219-1230.

[15] J. Arvo, Analytic methods for simulated light transport, Ph.D. thesis, Yale University, December 1995.

[16] S. Aydinli, H. Kaase, Measurement of luminous characteristics of daylighting materials, Tech. Rep. IEA SHCP TASK 21 / ECBCS ANNEX 29, Institute of Electronics and Lighting Technology, Technical University of Berlin, September 1999.

[17] J. Báni, Phototoxicity, photoirritation and photoallergy detection and assessment, in: P. Agache, P. Humbert (Eds.), Measuring the Skin, Springer-Verlag, Berlin, 2004, pp. 483–491.

[18] G. Baranoski, D. Eng, An investigation on sieve and detour effects affecting the interaction of collimated and diffuse infrared radiation (750 to 2500 nm) with plant leaves, IEEE Trans. Geosci. Remote Sens. 45 (8) (2007) 2593–2599.

[19] G. Baranoski, A. Krishnaswamy, An introduction to light interaction with human skin, Revista de Informática Teórica e Aplicada 11 (2004) 33–62.

[20] G. Baranoski, A. Krishnaswamy, B. Kimmel, An investigation on the use of data-driven scattering profiles in Monte Carlo simulations of ultraviolet light propagation in skin tissues, Phys. Med. Biol. 49 (2004) 4799–4809.

[21] G. Baranoski, A. Krishnaswamy, B. Kimmel, Increasing the predictability of tissue subsurface scattering simulations, The Vis. Comput. 21 (4) (2005) 265–278.

[22] G. Baranoski, J. Rokne, An algorithmic reflectance and transmittance model for plant tissue, Comput. Graph. Forum (EUROGRAPHICS Proceedings) 16 (3) (1997) 141–150.

[23] G. Baranoski, J. Rokne, Light Interaction with Plants: A Computer Graphics Perspective, Horwood Publishing, Chichester, UK, 2004.

[24] G. Baranoski, J. Rokne, Rendering plasma phenomena: applications and challenges, Comput. Graph. Forum 26 (4) (2007) 743–768.

[25] G. Baranoski, J. Rokne, G. Xu, Virtual spectrophotometric measurements for biologically and physically-based rendering, The Vis. Comput. 17 (8) (2001) 506–518.

[26] W. Barkas, Analysis of light scattered from a surface of low gloss into its specular and diffuse components, Proc. Phys. Soc. Lond. 51 (1939) 274–295.

[27] J. Barth, J. Cadet, J. Césarini, T. Fitzpatrick, A. McKinlay, M. Mutzhas, M. Pathak, M. Peak, D. Sliney, F. Urbach, CIE-134 collection in photobiology and photochemistry, in: TC6-26 report: Standardization of the Terms UV-A1, UV-A2 and UV-B, Commission International de L'Eclairage, 1999.

[28] A. Bashkatov, E. Genina, V. Kochubey, V. Tuchin, Optical properties of human skin, subcutaneous and mucous tissues in the wavelength range from 400 to 2000 nm, J. Phys. D Appl. Phys. 38 (2005) 2543–2555.

[29] I. Bell, G. Baranoski, More than RGB: moving toward spectral color reproduction, ACM SIGGRAPH Course Notes, San Diego, CA, July 2003 (Course 24).

[30] I. Bell, G. Baranoski, Reducing the dimensionality of plant spectral databases, IEEE Trans. Geosci. Remote Sens. 14 (3) (2004) 570–577.

[31] E. Bendit, D. Ross, A technique for obtaining the ultraviolet absorption spectrum of solid keratin, Appl. Spectrosc. 15 (4) (1961) 103–105.

[32] J. Bennett, Polarization, in: M. Bass, E. Stryland, D. Williams, W. Wolfe (Eds.), Handbook of Optics (Volume I: Fundamentals, Techniques, & Design), Optical Society of America, McGraw-Hill Inc., New York, 1995, pp. 5.1-5.30 (Chapter 5).

[33] E. Berry, Diffuse reflection of light from a matt surface, J. Opt. Soc. Am. 7 (8) (1923) 627-633.

[34] J. Blinn, Models of light reflection for computer synthesized images, Comput. Graph. (SIGGRAPH Proceedings) 11 (2) (1977) 192-198.

[35] C. Bohren, Colors of snow, frozen waterfalls, and icebergs, J. Opt. Soc. Am. 73 (12) (1983) 1646-1652.

[36] C. Bohren, Scattering by particles, in: M. Bass, E. Stryland, D. Williams, W. Wolfe (Eds.), Handbook of Optics (Volume I: Fundamentals, Techniques, & Design), Optical Society of America, McGraw-Hill Inc., New York, 1995, pp. 6.1-6.21 (Chapter 6).

[37] M. Bouguer, Traite d'optique sur la gradation de la lumiére, M. Ábbe de Lacaille, Paris, 1760.

[38] W. Bruls, J. van der Leun, Forward scattering properties of human epidermal layers, Photochem. Photobiol. 40 (1984) 231-242.

[39] R. Burden, J. Faires, Numerical Analysis, fifth ed., PWS-KENT Publishing Company, Boston, 1993.

[40] W. Butler, Absorption spectroscopy in vivo: theory and application, Annu. Rev. Plant Physiol. 15 (1964) 451-470.

[41] S. Chandrasekhar, Radiative Transfer, Dover Publications Inc., New York, 1960.

[42] M. Chedekel, Photophysics and photochemistry of melanin, in: M.C.L. Zeise, T. Fitzpatrick (Eds.), Melanin: Its Role in Human Photoprotection, Valdenmar Publishing Company, Overland Park, Kansas, 1995, pp. 11-22, 2223b.

[43] B. Chen, K. Stamnes, J. Stamnes, Validity of the diffusion approximation in bio-optical imaging, Appl. Opt. 40 (34) (2001) 6356-6336.

[44] T. Chen, G. Baranoski, K. Lin, Bulk scattering approximations for HeNe laser transmitted through paper, Opt. Express 16 (26) (2008) 21762-21771.

[45] W. Cheong, S. Prahl, A. Welch, A review of the optical properties of biological tissues, IEEE J. Quantum Electron. 26 (12) (1990) 2166-2185.

[46] E. Church, P. Takacs, Surface scattering, in: M. Bass, E. Stryland, D. Williams, W. Wolfe (Eds.), Handbook of Optics (Volume I: Fundamentals, Techniques, & Design), Optical Society of America, McGraw-Hill Inc., New York, 1995, pp. 7.1-7.14 (Chapter 7).

[47] D. Churmakov, I. Meglinsky, S. Piletsky, D. Greenhalgh, Analysis of skin tissues spatial fluorescence distribution by the Monte Carlo simulation, J. Phys. D Appl. Phys. 36 (2003) 1722-1728.

[48] CIE, Colorimetry Official Recommendations of the International Commission on Illumination, Commission Internationale de L'Eclairage (CIE), May 1970, CIE Colorimetry Committee (E-1.3.1).

[49] E. Claridge, S. Cotton, P. Hall, M. Moncrieff, From colour to tissue histology: physics based interpretation of images of pigmented lesions, in: 5th International Conference on Medical Image Computing and Computer Assisted Intervention – MICCAI 2002, Springer-Verlag, Berlin, 2002, pp. 730–738 (Part I).

[50] E. Claridge, S. Cotton, P. Hall, M. Moncrieft, From colour to tissue histology: physics based interpretation of images of pigmented skin lesions, Med. Image Anal. 7 (2003) 489–502.

[51] F. Clarke, D. Parry, Helmholtz reciprocity: its validity and application to reflectometry, Ltg. Res. Technol. 17 (1) (1985) 1–11.

[52] M. Cohen, J. Wallace, Radiosity and Realistic Image Synthesis, Academic Press Professional, Cambridge, 1993.

[53] T. Coleman, C. Van Loan, Handbook of Matrix Computations, SIAM Publications, Philadelphia, PA, 1988.

[54] D. Contini, F. Martelli, G. Zaccanti, Photon migration through a turbid slab described by a model based on diffusion aproximation. I. theory, Appl. Opt. 39 (19) (1997) 4587–4599.

[55] R. Cook, K. Torrance, A reflectance model for computer graphics, ACM Trans. Graph. 1 (1) (1982) 7–24.

[56] S. Cotton, E. Claridge, Developing a predictive model of skin colouring, in: SPIE, vol. 2708, Medical Imaging 1996, 1996, pp. 814–825.

[57] B. Crowther, Computer modeling of integrating spheres, Appl. Opt. 35 (30) (1996) 5880–5886.

[58] K. Dana, B. van Ginneken, S. Nayar, J. Koenderink, Reflectance and texture of real world surfaces, ACM Trans. Graph. 18 (1) (1999) 1–34.

[59] M. Darvin, I. Gersonde, M. Meinke, W. Sterry, J. Lademann, Non-invasive *in vivo* determination of the carotenoids beta-carotene and lycopene concentrations in the human skin using the Raman spectroscopic method, J. Phys. D Appl. Phys. 38 (2005) 2096–2700.

[60] D. Delpy, M. Cope, P. Zee, S. Wray, J. Wyatt, Estimation of the optical pathlength through tissue from direct flight measurment, Phys. Med. Biol. 33 (12) (1988) 1433–1442.

[61] E. d'Eon, D. Luebke, E. Enderton, Efficient rendering of human skin, in: Rendering Techniques 2007: 18th Eurographics Workshop on Rendering, 2007, pp. 147–157.

[62] D. Dickey, R. Moore, J. Tulip, Using radiance predicted by the P3-Approximation in a spherical geometry to predict tissue optical properties, in: P. Brouwer (Ed.), Clinical Lasers and Diagnostics, SPIE, vol. 4156, Bellingham, Washington, 2001, pp. 181–188.

[63] D. Dickey, R. Morre, D. Rayner, J. Tulip, Light dosimetry using P3 approximation, Phys. Med. Biol. 46 (2001) 2359–2370.

[64] B. Diffey, A mathematical model for ultraviolet optics in skin, Phys. Med. Biol. 28 (6) (1983) 647–657.

[65] H. Ding, J. Lu, W. Wooden, P. Kragel, X. Hu, Refractive indices of human skin tissues at eight wavelengths and estimated dispersion relationships between 300 and 1600 nm, Phys. Med. Biol. 51 (2006) 1479–1489.

[66] M. Doi, S. Tominaga, Spectral estimation of human skin color using the Kubelka-Munk theory, in: R. Eschbach, G.C. Marcu (Eds.), SPIE/IS&T Electronic Imaging, SPIE, vol. 5008, Bellingham, Washington, 2003, pp. 221–228.

[67] J. Dongarra, J. Bunch, C. Moler, G. Stewart, LINPACK Users' Guides, SIAM Publications, Philadelphia, PA, 1979.

[68] C. Donner, H. Jensen, Light diffusion in multi-layered translucent materials, ACM Trans. Graph. 24 (3) (2005) 1032–1039.

[69] C. Donner, H. Jensen, A spectral BSSRDF for shading human skin, in: Rendering Techniques 2006: 17th Eurographics Workshop on Rendering, June 2006, pp. 409–418.

[70] C. Donner, H. Jensen, Rendering translucent materials using photon diffusion, in: Rendering Techniques 2007: 18th Eurographics Workshop on Rendering, June 2007, pp. 234–251.

[71] C. Donner, T. Weyrich, E. d'Eon, R. Ramamoorthi, S. Rusinkiewicz, A layered, heterogeneous reflectance model for acquiring and rendering human skin, ACM Trans. Graph. 27 (5) (2008) 1–12.

[72] R. Doornbos, R. Lang, M. Aalders, F. Cross, H. Sterenborg, The determination of *in vivo* human tissue optical properties and absolute chromophore concentrations using spatially resolved steady-state diffuse reflectance spectroscopy, Phys. Med. Biol. 44 (1999) 967–981.

[73] J. Dorsey, H. Rushmeier, F. Sillion, Digital Modeling of Material Appearance, Morgan Kaufmann/Elsevier, Amsterdam, Netherlands, 2007.

[74] A. Dunn, R. Richards-Kortum, Three-dimensional computation of light scattering from cells, IEEE Sel. Top. Quantum Electron. 2 (1996) 898–905.

[75] P. Dutré, K. Bala, P. Bekaert, Advanced Global Illumination, second ed., AK Peters Ltd., Wellesley, Massachusetts, 2006.

[76] Z.B.E. Anderson, C. Bischof, S. Blackford, J. Demmel, J. Dongarra, J.D. Croz, A. Greenbaum, S. Hammarling, A. McKenney, D. Sorensen, LAPACK Users' Guides, third ed., SIAM Publications, Philadelphia, PA, 1999.

[77] A. S. E284-91C, Standard terminology of appearance, in: L. Wolff, S. Shafer, G. Healey (Eds.), Physics-Based Vision Principles and Practice: Radiometry, Jones and Bartlett Publishers, Boston, 1992, pp. 146–161.

[78] W. Eagan, T. Hilgeman, Optical Properties of Inhomogeneous Materials, Academic Press, New York, 1979.

[79] G. Eason, A. Veitch, R. Nisbet, F. Turnbul, The theory of backscattering of light by blood, J. Phys. 11 (1978) 1463–1479.

[80] E. Claridge, S. Preece, An inverse method for the recovery of tissue parameters from colour images, in: C. Taylor, J. Noble (Eds.), Information Processing in Medical Imaging (IPMI), Springer, Berlin, 2003, pp. 306–317, LNCS 2732.

[81] D. Eng, G. Baranoski, The application of photoacoustic spectral data to the modeling of leaf optical properties in the visible range, IEEE Trans. Geosci. Remote Sens. 45 (12) (2007) 2593–2599.

[82] M. Everett, E. Yeargers, R. Sayre, R. Olsen, Penetration of epidermis by ultraviolet rays, Photochem. Photobiol. 5 (1966) 533–542.

[83] T. Farell, M. Patterson, B. Wilson, A diffusion theory model of spatially resolved, steady-state diffuse reflectance for the noninvasive determination of tissue optical properties in vivo, Med. Phys. 19 (1992) 879–888.

[84] P. Farrant, Color in Nature: A Visual and Scientific Exploration, Blandford Press, London, 1999.

[85] Y. Feldman, A. Puzenko, P. Ishai, A. Caduff, I. Davidovich, F. Sakran, A. Agranat, The electromagnetic response of human skin in the millimetre and submillimetre wave range, Phys. Med. Biol. 54 (2009) 3341–3363.

[86] R. Feynman, QED: The Strange Theory of Light and Matter, Princeton University Press, Princeton, New Jersey, 1985.

[87] R. Feynman, R. Leighton, M. Sands, The Feynman Lectures on Physics, vol. I, Addison-Wesley Publishing Company, Reading, Massachusetts, 1964.

[88] R. Feynman, R. Leighton, M. Sands, The Feynman Lectures on Physics, vol. II, Addison-Wesley Publishing Company, Reading, Massachusetts, 1964.

[89] T. Fitzpatrick, Soleil et peau, J. Med. Esthet. 2 (1975) 33–34.

[90] T. Fitzpatrick, J. Bolognia, Human melanin pigmentation: role in pathogenesis of cutaneous melanoma, in: M.C.L. Zeise, T. Fitzpatrick (Eds.), Melanin: Its Role in Human Photoprotection, Valdenmar Publishing Company, Overland Park, Kansas, 1995, pp. 177–182.

[91] R. Flewelling, Noninvasive optical monitoring, in: J. Bronzino (Ed.), The Biomedical Engineering Handbook, IEEE Press, Boca Raton, Florida, 1995, pp. 1346–1356 (Chapter 88).

[92] S.T. Flock, M.S. Patterson, B.C. Wilson, D.R. Wyman, Monte Carlo modeling of light propagation in highly scattering tissues - I: Model predictions and comparison with diffusion theory, IEEE Trans. Biomed. Eng. 36 (12) (1989) 1162–1168.

[93] J. Foley, A. van Dam, S. Feiner, J. Hughes, Computer Graphics: Principles and Practice, second ed., Addison-Wesley Publishing Company, Reading, Massachusetts, 1990.

[94] S. Foo, A gonioreflectometer for measuring the bidirectional reflectance of material for use in illumination computation, Master's thesis, Cornell University, August 1997.

[95] A. Fournier, From local to global illumination and back, in: P.M. Hanrahan, W. Purgathofer (Eds.), Rendering Techniques '95 (Proceedings of the Sixth Eurographics Rendering Workshop), Springer-Verlag, Dublin, 1995, pp. 127–136.

[96] R. Fretterd, R. Longini, Diffusion dipole source, J. Opt. Soc. Am. 63 (3) (1973) 336–337.

[97] L. Fukshansky, Optical properties of plants, in: H. Smith (Ed.), Plants and the Daylight Spectrum, Academic Press, London, 1981, pp. 21–40.

[98] K. Furutso, Diffusion equation derived from space-time transport equation, J. Opt. Soc. Am. 70 (1980) 360.

[99] C. Gerald, P. Wheatley, Applied Numerical Analysis, sixth ed., Addison-Wesley, Reading, Massachusetts, 1997.

[100] A. Glassner, Principles of Digital Image Synthesis, Morgan Kaufmann Publishers Inc., San Francisco, 1995.

[101] G. Golub, C.V. Loan, Matrix Computations, second ed., John Hopkins University Press, Baltimore, 1989.

[102] Y. Govaerts, S. Jacquemoud, M. Verstraete, S. Ustin, Three-dimensional radiation transfer modeling in a dycotyledon leaf, Appl. Opt. 35 (33) (1996) 6585–6598.

[103] K. Govidan, J. Smith, L. Knowles, A. Harvey, P. Townsend, J. Kenealy, Assessment of nurse-led screening of pigmented lesions using SIAscope, J. Plast. Reconstr. Aesthet. Surg. 60 (2007) 639–645.

[104] L. Grant, C. Daughtry, V. Vanderbilt, Polarized and specular reflectance variance with leaf surface features, Physiol. Plant. 88 (1) (1993) 1–9.

[105] R. Groenhuis, H. Fewerda, J. Bosch, Scattering and absorption of turbid materials determined from reflection measurements. 1: Theory, Appl. Opt. 22 (16) (1983) 2456–2462.

[106] J. Guillod, P. Schmid, Dermoscopy, in: P. Agache, P. Humbert (Eds.), Measuring the Skin, Springer-Verlag, Berlin, 2004, pp. 60–73.

[107] R. Hall, Comparing spectral color computation methods, IEEE Comput. Graph. Appl. 19 (4) (1999) 36–45.

[108] J. Hammerley, D. Handscomb, Monte Carlo Methods, Wiley, New York, 1964.

[109] M. Haniffa, J. Lloyd, C. Lawrence, The use of a spectrophotometric intracutaneous analysis device in the real-time diagnosis of melanoma in the setting of a melanoma screening clinic, Br. J. Dermatol. 156 (2007) 1350–1352.

[110] P. Hanrahan, W. Krueger, Reflection from layered surfaces due to subsurface scattering, in: SIGGRAPH, Annual Conference Series, 1993, pp. 165–174.

[111] E. Hecht, A. Zajac, Optics, Addison-Wesley, Reading, Massachusetts, 1974.

[112] P. Heckbert, Writing a ray tracer, in: A. Glassner (Ed.), An Introduction to Ray Tracing, Academic Press, San Diego, CA, 1989.

[113] L. Henyey, J. Greenstein, Diffuse radiation in the galaxy, Astrophys. J. 93 (1941) 70–83.

[114] A. Hielscher, R. Alcouffe, R. Barbour, Comparison of finite-difference transport and diffusion calculations for photon migration in homogeneous tissues, Phys. Med. Biol. 43 (1998) 1285–1302.

[115] R. Hirko, R. Fretterd, R. Longini, Application of the diffusion dipole to modelling the optical characteristics of blood, Med. Biol. Eng. 3 (1975) 192–195.

[116] C. Hourdakis, A. Perris, A Monte Carlo estimation of tissue optical properties for use in laser dosimetry, Phys. Med. Biol. 40 (1995) 351–364.

[117] E. Hudson, M. Stringer, F. Cairnduff, D. Ash, M. Smith, The optical properties of skin tumors measured during superficial photodynamic therapy, Lasers Med. Sci. 9 (1994) 99–103.

[118] R. Hunt, Measuring Colour, second ed., Ellis Horwood Limited, London, 1991.

[119] R. Hunter, R. Harold, The Measurement of Appearance, second ed., John Wiley & Sons, New York, 1987.

[120] F. Imai, Color reproduction of facial pattern and endoscopic image based on color appearance models, Ph.D. thesis, Graduate School of Science and Technology, Chiba University, Japan, December 1996.

[121] F. Imai, N. Tsumura, H. Haneishi, Y. Miyake, Principal component analysis of skin color and its application to colorimetric reproduction on CRT display and hardcopy, J. Imaging Sci. Technol. 40 (1996) 422–430.

[122] D. Immel, M. Cohen, D. Greenberg, A radiosity method for non-diffuse environments, Comput. Graph. (SIGGRAPH Proceedings) 20 (4) (1986) 133–142.

[123] A. Ishimaru, Wave Propagation and Scattering in Random Media, vol. 1, second ed., IEEE Press, New York, 1978.

[124] S. Jacquemoud, S. Ustin, Leaf optical properties: a state of the art, in: 8th International Symposium of Physical Measurements & Signatures in Remote Sensing, CNES, Aussois, France, 2001, pp. 223–332.

[125] S. Jacquemoud, S. Ustin, J. Verdebout, G. Schmuck, G. Andreoli, B. Hosgood, Estimating leaf biochemistry using prospect leaf optical properties model, Rem. Sens. Environ. 56 (1996) 194–202.

[126] S. Jacques, Origins of tissue optical properties in the uva visible and nir regions, OSA TOPS on Advances in Optical Imaging and Photon Migration 2 (1996) 364–369.

[127] S. Jacques, Optical absorption of melanin, Tech. Rep. Oregon Medical Laser Center, Portland, Oregon, 2001.

[128] S. Jacques, C. Alter, S. Prahl, Angular dependence of He-Ne laser light scattering by human dermis, Laser. Life Sci. 1 (4) (1987) 309–333.

[129] S. Jacques, D. McAuliffe, The melanosome: threshold temperature for explosive vaporization and internal absorption coefficient during pulsed laser irradiation, Photochem. Photobiol. 53 (6) (1991) 769–775.

[130] H. Jensen, J. Buhler, A rapid hierarchical rendering technique for translucent materials, in: SIGGRAPH, Annual Conference Series, July 2002, pp. 576–581.

[131] H. Jensen, J. Buhler, Digital face cloning, in: M.C.L. Zeise, T. Fitzpatrick (Eds.), SIGGRAPH 2003 Technical Sketches, Valdenmar Publishing Company, Overland Park, Kansas, 2003, pp. 11–22, 2223b.

[132] H. Jensen, S. Marschner, M. Levoy, P. Hanrahan, A practical model for subsurface light transport, in: SIGGRAPH, Annual Conference Series, August 2001, pp. 511–518.

[133] K. Jimbow, K. Reszka, S. Schmitz, T. Salopek, P. Thomas, Distribution of ue- and pheomelanins in human skin and melanocytic tumors, and their photoprotective vs phototoxic properties, in: M.C.L. Zeise, T. Fitzpatrick (Eds.), Melanin:

Its Role in Human Photoprotection, Valdenmar Publishing Company, Overland Park, Kansas, 1995, pp. 165–175.

[134] Z. Jin, K. Stammes, Radiative transfer in nonuniformly refracting layered media: atmosphere-ocean system, Appl. Opt. 33 (3) (1994) 431–442.

[135] D. Judd, G. Wyszecki, Color in Business, Science and Industry, third ed., John Wiley & Sons, New York, 1975.

[136] J. Kajiya, The rendering equation, Comput. Graph. (SIGGRAPH Proceedings) 20 (4) (1986) 143–150.

[137] M. Kalos, P. Whitlock, Monte Carlo Methods, vol. I: Basics, John Wiley & Sons, New York, 1986.

[138] G. Kattawar, A three-parameter analytic phase function for multiple scattering calculations, J. Quant. Spectrosc. Radiat. Transf. 15 (1975) 839–849.

[139] G. Kelfkens, J. van der Leun, Skin temperature changes after irradiation with UVB or UVC: implications for the mechanism underlying ultraviolet erythema, Phys. Med. Biol. 34 (5) (1989) 599–608.

[140] B. Kimmel, G. Baranoski, A novel approach for simulating light interaction with particulate materials: application to the modeling of sand spectral properties, Opt. Express 15 (15) (2007) 9755–9777.

[141] B. Kimmel, G. Baranoski, A compact framework to efficiently represent the reflectance of sand samples, IEEE Trans. Geosci. Remote Sens. 47 (11) (2009) 3625–3629.

[142] M. Kobayashi, Y. Ito, N. Sakauchi, I. Oda, I. Konishi, Y. Tsunazawa, Analysis of nonlinear relation for skin hemoglobin, Opt. Express 9 (3) (2001) 802–812.

[143] C. Kolb, Rayshade User's Guide and Reference Manual, Princeton University, Princeton, New Jersey, January 1992.

[144] N. Kollias, The spectroscopy of human melanin pigmentation, in: M.C.L. Zeise, T. Fitzpatrick (Eds.), Melanin: Its Role in Human Photoprotection, Valdenmar Publishing Company, Overland Park, Kansas, 1995, pp. 31–38.

[145] K. Kölmel, B. Sennhenn, K. Giese, Investigation of skin by ultraviolet remittance spectroscopy, Br. J. Dermatol. 122 (1990) 209–216.

[146] A. Krishnaswamy, BioSpec: a biophysically-based spectral model of light interaction with human skin, Master's thesis, School of Computer Science, University of Waterloo, Waterloo, Ontario, Canada, 2005.

[147] A. Krishnaswamy, G. Baranoski, A biophysically-based spectral model of light interaction with human skin, Comput. Graph. Forum (EUROGRAPHICS Proceedings) 23 (3) (2004) 331–340.

[148] A. Krishnaswamy, G. Baranoski, Combining a shared-memory high performance computer and a heterogeneous cluster for the simulation of light interaction with human skin, in: P.N.J. Gaudiot, M.L. Pilla, S. Song (Eds.), 16th Symposium on Computer Architecture and High Performance Computing, IEEE Computer Society, Washington, 2004, pp. 166–171.

[149] A. Krishnaswamy, G. Baranoski, J.G. Rokne, Improving the reliability/cost ratio of goniophotometric measurements, J. Graph. Tool. 9 (3) (2004) 31–51.

[150] P. Kubelka, F. Munk, Ein beitrag zur optik der farbanstriche, Zurich Tech. Physik 12 (1931) 543.

[151] E. Lafortune, Mathematical models and Monte Carlo algorithms for physically based rendering, Ph.D. thesis, Department of Computer Science, Faculty of Engineering, Katholieke Universiteit Leuven, February 1996.

[152] R. Lee, M. Mathews-Roth, M. Pathak, J. Parrish, The detection of carotenoid pigments in human skin, J. Invest. Dermatol. 64 (1975) 175–177.

[153] J. Lenoble, Atmospheric Radiative Transfer, A. Deepak Publishing, Hampton, Virginia, 1993.

[154] D. Leroy, Skin photoprotection function, in: P. Agache, P. Humbert (Eds.), Measuring the Skin, Springer-Verlag, Berlin, 2004, pp. 471–482.

[155] J. Lévêque, B. Querleux, SkinChip, a new tool for investigating the skin surface *in vivo*, Skin Res. Technol. 9 (2003) 343–347.

[156] G. Lewis, The conservation of photons, Nature 2981 (118) (1926) 874–875.

[157] S. Li, Biologic biomaterials: tissue-derived biomaterials (collagen), in: J. Park, J. Bronzano (Eds.), Biomaterials Principles and Applications, CRC Press, Boca Raton, Florida, 2003, pp. 117–139.

[158] C. Lilley, F. Lin, W. Hewitt, T. Howard, Colour in Computer Graphics, ITTI Computer graphics and Visualisation, Manchester Computing Centre, The University of Manchester, Manchester, England, 1993.

[159] R. Longhurst, Geometrical and Physical Optics, third ed., Longman Group Limited, London, 1973.

[160] A. Lovell, J. Hebden, J. Goldstone, M. Cope, Determination of the transport scattering coefficient of red blood cells, in: B. Chance, R.R. Alfonso, B.J. Tromberg (Eds.), Optical Tomography and Spectroscopy of Tissue III, SPIE, vol. 3597, Bellingham, Washington, 1999, pp. 121–128.

[161] G. Lucassen, P. Caspers, G. Puppels, Infrared and Raman spectroscopy of human skin *in vivo*, in: V. Tuchin (Ed.), Handbook of Optical Biomedical Diagnostics, SPIE Press, Bellingham, Washington, 2002, pp. 787–824.

[162] Q. Ma, A. Nishimura, P. Phu, Y. Kuga, Transmission, reflection and depolarization of an optical wave for a single leaf, IEEE Trans. Geosci. Remote Sens. 28 (5) (1990) 865–872.

[163] D. MacAdam, Color Measurements Theme and Variations, Springer Verlag, Berlin, 1981.

[164] S. Marschner, S.H. Westin, E. Lafortune, K. Torrance, D. Greenberg, Image-based BRDF measurement, Tech. Rep. PCG-99-1, Program of Computer Graphics, Cornell University, New York, January 1999.

[165] S. Marschner, S.H. Westin, E. Lafortune, K. Torrance, D. Greenberg, Image-based BRDF measurement including human skin, in: D. Lischinski, G.W. Larson (Eds.), Rendering Techniques 1999 (Proceedings of the 10th Eurographics Rendering Workshop), Springer-Verlag, Granada, 1999, pp. 119–130.

[166] S. Marschner, S.H. Westin, E. Lafortune, K. Torrance, D. Greenberg, Reflectance measurements of human skin, Tech. Rep. PCG-99-2, Program of Computer Graphics, Cornell University, New York, January 1999.

[167] P. Matts, S. Cotton, Spectrophotometric intracutaneous analysis (SIAscopy), in: A. Barel, M. Paye, H. Maibach (Eds.), Handbook of Cosmetic Science and Technology, Informa Health Care, New York, 2009, pp. 275–281.

[168] E. McCartney, Optics of the Atmosphere: Scattering by Molecules and Particles, John Wiley & Sons Inc., New York, 1976.

[169] R. McCluney, Introduction to Radiometry and Photometry, Artech House Inc., Boston, 1994.

[170] I. Meglinsky, S. Matcher, Modelling the sampling volume for skin blood oxygenation, Med. Biol. Eng. Comput. 39 (2001) 44–49.

[171] I. Meglinsky, S. Matcher, Quantitative assessment of skinlayers absorption and skin reflectance spectra simulation in the visible and near-infrared spectral regions, Physiol. Meas. 23 (2002) 741–753.

[172] I. Meglinsky, S. Matcher, Computer simulation of the skin reflectance spectra, Comput. Methods Programs Biomed. 70 (2003) 179–186.

[173] D. Menzel, Selected Papers on the Transfer of Radiation, Dover Publications, New York, 1966.

[174] N. Metropolis, S. Ulam, The Monte Carlo method, J. Am. Stat. Assoc. 44 (247) (1949) 335–341.

[175] J. Meyer-Arendt, Introduction to Modern and Classical Optics, Prentice-Hall, New Jersey, 1984.

[176] M.F. Yang, V. Tuchin, A. Yaroslavsky, Principles of light-skin interactions, in: E. Baron (Ed.), Light-Based Therapies for Skin of Color, Springer-Verlag, London, 2009, pp. 1–44.

[177] M. Moncrieff, S. Cotton, E. Claridge, P. Hall, Spectrophotometric intracutaneous analysis: a new technique for imaging pigmented skin lesions, Br. J. Dermatol. 146 (2002) 448–457.

[178] J. Mourant, J. Freyer, A. Hielscher, A. Eick, D. Shen, T. Johnson, Mechanisms of light scattering from biological cells relevant to noninvasive optical-tissue diagnostics, Appl. Opt. 37 (16) (1998) 3586–3593.

[179] H. Nakai, Y. Manabe, S. Inokuchi, Simulation analysis of spectral distributions of human skin, in: 14th International Conference on Pattern Recognition, 1998, pp. 1065–1067.

[180] C. Ng, L. Li, A multi-layered reflection model of natural human skin, in: Computer Graphics International 2001, Hong Kong, July 2001, pp. 249–256.

[181] F. Nicodemus, J. Richmond, J. Hsia, I. Ginsberg, T. Limperis, Geometrical considerations and nomenclature for reflectance, in: L. Wolff, S. Shafer, G. Healey (Eds.), Physics-Based Vision Principles and Practice: Radiometry, Jones and Bartlett Publishers, Boston, 1992, pp. 94–145.

[182] K. Nielsen, L. Zhao, J. Stamnes, K. Stamnes, J. Moan, Reflectance spectra of pigmented and nonpigmented skin in the UV spectral region, Photochem. Photobiol. 80 (2004) 450–455.

[183] K. Nielsen, L. Zhao, J. Stamnes, K. Stamnes, J. Moan, The importance of the depth distribution of melanin in skin for DNA protection and other photobiological processes, J. Photochem. Photobiol. B, Biol. 82 (2006) 194–198.

[184] S. Nilsson, Skin temperature over an artificial heat source implanted in man, Phys. Med. Biol. 20 (3) (1975) 366–383.

[185] M. Nischik, C. Forster, Analysis of skin erythema using true-color images, IEEE Trans. Med. Imaging 16 (6) (1997) 711–716.

[186] I. Nishidate, Y. Aizu, H. Mishina, Estimation of melanin and hemoglobin in skin tissue using multiple regression analysis aided by Monte Carlo simulation, J. Biomed. Opt. 9 (4) (2004) 700–710.

[187] A. Nunez, M. Mendehall, Detection of human skin in near infrared hyperspectral imagery, in: International Geoscience and Remote Sensing Symposium – IGARSS '08, 2008, pp. II-621–624.

[188] N. Ohta, A. Robertson, Colorimetry Fundamentals and Applications, John Wiley & Sons, New York, 1982.

[189] S. Orchard, Reflection and transmission of light by diffusing suspensions, J. Opt. Soc. Am. 59 (1969) 1584–1597.

[190] R. Overhem, D. Wagner, Light and Color, John Wiley & Sons, New York, 1982.

[191] K. Palmer, D. Williams, Optical properties of water in the near infrared, J. Opt. Soc. Am. 64 (8) (1974) 1107–1110.

[192] D. Parsad, K. Wakamatsu, A. Kanwar, B. Kumar, S. Ito, Eumelanin and phaeomelanin contents of depigmented and repigmented skin in vitiligo patients, Br. J. Dermatol. 149 (2003) 624–626.

[193] M. Pathak, Functions of melanin and protection by melanin, in: M.C.L. Zeise, T. Fitzpatrick (Eds.), Melanin: Its Role in Human Photoprotection, Valdenmar Publishing Company, Overland Park, Kansas, 1995, pp. 125–134.

[194] A. Pearce, K. Sung, Maya software rendering: a technical overview, Tech. Rep. AP-M-SWR-01, Alias—Wavefront, Toronto, Canada, 1998.

[195] A. Petitjean, P. Humbert, S. Mac-Mary, J. Sainthillier, Skin radiance measurement, in: A. Barel, M. Paye, H. Maibach (Eds.), Handbook of Cosmetic Science and Technology, Informa Health Care, New York, 2009, pp. 407–414.

[196] E. Pickwell, B. Cole, A. Fitzgerald, M. Pepper, V. Wallace, *In vivo* study of human skin using pulsed terahertz radiation, Phys. Med. Biol. 49 (2004) 1595–1607.

[197] R. Pope, E. Fry, Absorption spectrum (380–700 nm) of pure water. II. integrating cavity measurements, Appl. Opt. 36 (33) (1997) 8710–8723.

[198] A. Popov, A. Priezzhev, Laser pulse in turbid media: Monte Carlo simulation and comparison with experiment, in: V. Tuchin (Ed.), Saratov Fall Meeting 2002: Optical Technologies in Biophysics and Medicine IV, SPIE vol. 5068, Bellingham, Washington, 2003, pp. 299–308.

[199] A. Popov, A. Priezzhev, J. Lademann, R. Myllylä, TiO_2 nanoparticles as an effective UV-B radiation skin-protective compound in sunscreens, J. Phys. D Appl. Phys. 38 (2005) 2564–2570.

[200] P. Poulin, A. Fournier, A model for anisotropic reflection, Comput. Graph. (SIGGRAPH Proceedings) 24 (4) (1990) 273–282.

[201] S. Prahl, Light transport in tissue, Ph.D. thesis, The University of Texas at Austin, Texas, December 1988.

[202] S. Prahl, Optical absorption of hemoglobin, Tech. Rep. Oregon Medical Laser Center, Portland, Oregon, 1999.

[203] S. Prahl, PhotochemCAD spectra by category, Tech. Rep. Oregon Medical Laser Center, Portland, Oregon, 2001.

[204] S. Prahl, M. Keijzer, S. Jacques, A. Welch, A Monte Carlo model of light propagation in tissue, SPIE Institute Series IS 5 (1989) 102–111.

[205] S. Prahl, M. van Gemert, A. Welch, Determining the optical properties of turbid media using the adding-doubling method, Appl. Opt. 32 (4) (1993) 559–568.

[206] A. Pravdin, S. Chernova, T. Papazoglou, V. Tuchin, L. Wang, Tissue phantoms, in: V. Tuchin (Ed.), Handbook of Optical Biomedical Diagnostics, SPIE Press, Bellingham, Washington, 2002, pp. 311–352.

[207] R. Preisendorfer, Radiative Transfer on Discrete Spaces, Pergamon, New York, 1965.

[208] E. Questel, Y. Gall, Photobiological assessment of sunscreens, in: P. Agache, P. Humbert (Eds.), Measuring the Skin, Springer-Verlag, Berlin, 2004, pp. 492–505.

[209] L. Reynolds, C. Johnson, A. Ishimaru, Diffuse reflectance from a finite blood medium: applications to the modeling of fiber optics catheters, Appl. Opt. 15 (9) (1976) 2059–2067.

[210] J. Rodriguez, I. Yaroslavsky, A. Yaroslavsky, H. Battarbee, V. Tuchin, Time-resolved imaging in diffusive media, in: V. Tuchin (Ed.), Handbook of Optical Biomedical Diagnostics, SPIE Press, Bellingham, Washington, 2002, pp. 357–404.

[211] B. Rolinsky, H. Küster, B. Ugele, R. Gruber, K. Horn, Total bilirubin measurement by photometry on a blood gas analyser: potential for use in neonatal testing at point of care, Clin. Chem. 47 (10) (2001) 1845–1847.

[212] A. Rosencwaig, Photoacoustics and Photoacoustics Spectroscopy, Wiley, New York, 1980.

[213] W. Ruhle, A. Wild, The intensification of absorbance changes in leaves by light-dispersion: differences between high-light and low-light leaves, Planta 146 (1979) 551–557.

[214] I. Saidi, Transcutaneous optical measurement of hyperbilirubinemia in neonates, Ph.D. thesis, Rice University, Houston, Texas, 1994.

[215] H. Samet, The quadtree and related hierarchical data structures, ACM Computing Surveys 16 (2) (1984) 187–260.

[216] D. Sardar, L. Levy, Optical properties of whole blood, Lasers Med. Sci. 13 (1998) 106–111.

[217] J. Schmitt, G. Zhou, E. Walker, R. Wall, Multilayer model of photon diffusion in skin, J. Opt. Soc. Am. 7 (11) (1990) 2141–2153.

[218] A. Schuster, Radiation through foggy atmosphere, Astrophys. J. 21 (1) (1905) 1–22.

[219] F. Sears, M. Zemansky, H. Young, College Physics Part I Mechanics, Heat and Sound, fourth ed., Addison-Wesley Publishing Company, Reading, Massachusetts, 1974.

[220] H. Seeliger, The photometry of diffusely reflecting surfaces, Koniglich Bayerische Akademie der Wissenschaften 18 (1888) 201–248 (in German).

[221] J. Shawe-Taylor, N. Cristianini, Kernel Methods for Pattern Analysis, University Press, Cambridge, 2004.

[222] T. Shi, C. DiMarzio, Multispectral method for skin imaging: development and validation, Appl. Opt. 46 (36) (2007) 8619–8626.

[223] M. Shimada, Y. Yamada, M. Itoh, M. Takahashi, T. Yatagai, Explanation of human skin color by multiple linear regression analysis based on the modified Lambert-Beer law, Opt. Rev. 7 (3) (2000) 348–352.

[224] M. Shimada, Y. Yamada, M. Itoh, T. Yatagai, Melanin and blood concentration in human skin studied by multiple regression analysis: assessment by Monte Carlo simulation, Phys. Med. Biol. 46 (2001) 2397–2406.

[225] P. Shirley, Physically based lighting for computer graphics, Ph.D. thesis, Dept. of Computer Science, University of Illinois, November 1990.

[226] P. Shirley, A ray tracing method for illumination calculation in diffuse-specular scenes, in: Graphics Interface, Canadian Information Processing Society, Toronto, 1990, pp. 205–212.

[227] P. Shirley, Nonuniform random points via warping, in: D. Kirk (Ed.), Graphics Gems III, Academic Press, Boston, 1992, pp. 80–83.

[228] A. Shiryaev, Probability, second ed., Springer-Verlag, New York, 1996.

[229] C. Simpson, M. Kohl, M. Essenpreis, M. Cope, Near-infrared optical properties of *ex-vivo* human skin and subcutaneous tissues measured using the Monte Carlo inversion technique, Phys. Med. Biol. 43 (1998) 2465–2478.

[230] Y. Sinichkin, N. Kolias, G. Zonios, S. Utz, V. Tuchin, L. Wang, Reflectance and fluorescence spectroscopy of human skin *in vivo*, in: V. Tuchin (Ed.), Handbook of Optical Biomedical Diagnostics, SPIE Press, Bellingham, Washington, 2002, pp. 725–786.

[231] H. Smith, D. Morgan, The spectral characteristics of the visible radiation incident upon the surface of the Earth, in: H. Smith (Ed.), Plants and the Daylight Spectrum, Academic Press, London, 1981, pp. 3–20.

[232] J. Stam, An illumination model for a skin layer bounded by rough surfaces, in: P.M. Hanrahan, W. Purgathofer (Eds.), Rendering Techniques 2001 (Proceedings of the 12th Eurographics Rendering Workshop), Springer-Verlag, London, 2001, pp. 39–52.

[233] K. Stamnes, P. Conklin, A new multi-layer discrete ordinate approach to radiative transfer in vertically inhomogeneous atmospheres, J. Quantum Spectrosc. Radiat. Transf. 31 (3) (1984) 273–282.

[234] K. Stamnes, S. Tsay, W. Wiscombe, K. Jayaweera, Numerically stable algorithm for discrete-ordinate-method radiative transfer in multiple scattering and emitting layered media, Appl. Opt. 27 (12) (1988) 2502–2509.

[235] K. Stanzl, L. Zastrow, Melanin: an effective photoprotectant against UV-A rays, in: M.C.L. Zeise, T. Fitzpatrick (Eds.), Melanin: Its Role in Human Photoprotection, Valdenmar Publishing Company, Overland Park, Kansas, 1995, pp. 59–63.

[236] W. Star, Comparing the P3-Approximation with diffusion theory and with Monte Carlo calculations of light propagation in a slab geometry, in: SPIE Institute Series 5: Dosimetry of Laser Radiation in Medicine and Biology, SPIE-The International Society for Optical Engineering, Bellingham, Washington, 1989, pp. 146–154.

[237] W. Star, Light dosimetry *in vivo*, Phys. Med. Biol. 42 (1997) 763–787.

[238] J. Steinke, A. Shepherd, Diffusion model of the optical absorbance of whole blood, J. Opt. Soc. Am. 5 (6) (1988) 813–822.

[239] W. Sthal, H. Sies, Carotenoids in systemic protecion against sunburns, in: N. Krisnky, S. Mayne, H. Sies (Eds.), Carotenoids in Health and Disease, CRC Press, Boca Raton, Florida, 2004, pp. 491–502.

[240] M. Stone, A Field Guide to Digital Color, AK Peters, Natick, MA, 2003.

[241] M. Störring, Computer vision and human skin color, Ph.D. thesis, Faculty of Engineering and Science, Aalborg University, Denmark, 2004.

[242] J. Strutt, On the transmission of light through an atmosphere containing many small particles in suspension, and on the origin of the blue of the sky, Philos. Mag. 47 (1899) 375–384.

[243] Q. Sun, M. Fairchild, Statistical characterization of spectral reflectances in spectral imaging of human portraiture, in: Ninth Color Imaging Conference: Color Science and Engineering, 2001, pp. 73–79.

[244] P. Talreja, G. Kasting, N. Kleene, W. Pickens, T. Wang, Visualization of the lipid barrier and measurement of lipid pathlength in human stratum corneum, AAPS PharmSCi 3 (2) (2001) 1–9.

[245] H. Tehrani, J. Walls, S. Cotton, E. Sassoon, P. Hall, Spectrophotometric intracutaneous analysis in the diagnosis of basal cell carcinoma: a pilot study, Int. J. Dermatol. 46 (2007) 371–375.

[246] J. Tessendorf, D. Wilson, Impact of multiple scattering on simulated infrared cloud scene images, in: P. Christensen, D. Cohen-Or (Eds.), SPIE. Characterization and Propagation of Sources and Backgrounds, Bellingham, Washington, 1994, pp. 75–84, 2223b.

[247] N. Thalmann, P. Kalra, J. Lévêque, R. Bazin, D. Batisse, B. Querleux, A computational skin model: fold and wrinkle formation, IEEE Trans. Inf. Technol. Biomed. 6 (4) (2002) 317–323.

[248] A. Thody, E. Higgins, K. Wakamatsu, S. Ito, S. Burchill, J. Marks, Pheomelanin as well as eumelanin is present in human dermis, J. Invest. Dermatology 97 (1991) 340–344.

[249] K. Torrance, E. Sparrow, Theory for off-specular reflection from roughened surfaces, J. Opt. Soc. Am. 57 (9) (1967) 1105–1114.

[250] W. Tropf, M. Thomas, T. Harris, Properties of crystals and glasses, in: M. Bass, E. Stryland, D. Williams, W. Wolfe (Eds.), Handbook of Optics (Volume II: Devices, Measurements, and Properties), Optical Society of America, McGraw-Hill Inc., New York, 1995, pp. 33.1–33.101 (Chapter 33).

[251] T. Trowbridge, K. Reitz, Average irregularity representation of a rough surface for ray reflection, J. Opt. Soc. Am. 65 (5) (1975) 531–536.

[252] T. Troy, S.N. Thennadil, Optical properties of human skin in NIR wavelength range of 1000–2000 nm, J. Biomed. Opt. 6 (2) (2001) 167–176.

[253] N. Tsumura, M. Kawabuchi, H. Haneishi, Y. Miyabe, Mapping pigmentation in human skin by multi- visible-spectral imaging by inverse optical scattering technique, in: IS&T/SID Eighth Color Imaging Conference, 2000, pp. 81–84.

[254] V. Tuchin, Tissue Optics: Light Scattering Methods and Instruments for Medical Diagnosis, The International Society for Optical Engineering, Bellingham, Washington, 2000.

[255] V. Tuchin, Optical clearing of tissues and blood using the immersion method, J. Phys. D Appl. Phys. 38 (2005) 2497–2518.

[256] V. Tuchin, S. Utz, I. Yaroslavsky, Tissue optics, light distribution, and spectroscopy, Opt. Eng. 33 (1994) 3178–3188.

[257] A. Uesugi, W. Irvine, Y. Kawata, Formation of absorption spectra by diffuse reflection from a semi-infinite planetary atmosphere, J. Quant. Spectrosc. Radiat. Transf. 11 (1971) 797–808.

[258] P. Urso, M. Lualdi, A. Colombo, M. Carrara, S. Tomatis, R. Marchesini, Skin and cutaneous melanocytic lesion simulation in biomedical optics with multilayered phantoms, Phys. Med. Biol. 52 (2007) N229–N239.

[259] J. Uspensky, Introduction to Mathematical Probability, McGraw-Hill, New York, 1937.

[260] H. van de Hulst, Multiple Light Scattering: Tables, Formulas, and Applications, vol. 1, Academic Press, New York, 1980.

[261] H. van de Hulst, Multiple Light Scattering: Tables, Formulas, and Applications, vol. 2, Academic Press, New York, 1980.

[262] H. van de Hulst, Light Scattering by Small Particles, second ed., Dover Publications Inc., New York, 1981.

[263] J. van der Leun, Ultraviolet erythema, Ph.D. thesis, University of Utrecht, The Netherlands, 1966.

[264] M. van Gemert, S. Jacques, H. Sterenborg, W. Star, Skin optics, IEEE Trans. Biomed. Eng. 36 (12) (1989) 1146–1154.

[265] M. van Gemert, W. Star, Relations between the Kubelka-Munk and the transport equation models for anisotropic scattering, Laser Life Sci. 1 (4) (1987) 287–298.

[266] M. van Gemert, A. Welch, W. Star, M. Motamedi, W. Cheong, Tissue optics for a slab geometry in diffusion approximation, Lasers Med. Sci. 2 (1987) 295–302.

[267] B. van Ginneken, M. Stavridi, J. Koenderink, Diffuse and specular reflectance from rough surfaces, Appl. Opt. 37 (1) (1998) 130–139.

[268] E. Veach, Robust monte carlo methods for light transport simulation, Ph.D. thesis, Stanford University, December 1997.

[269] J. Viator, J. Komadina, L. Svaasand, G. Aguilar, B. Choi, J. Nelson, A comparative study of photoacoustic and reflectance methods for determination of epidermal melanin content, J. Invest. Dermatol. 122 (2004) 1432–1439.

[270] M. Vrhel, R. Gershon, L. Iwan, Measurement and analysis of object reflectance spectra, Color Res. Appl. 19 (1) (1994) 4–9.

[271] S. Wan, R. Anderson, J. Parrish, Analytical modeling for the optical properties of the skin with *in vitro* and *in vivo* applications, Photochem. Photobiol. 34 (1981) 493–499.

[272] L. Wang, Rapid modeling of diffuse reflectance in turbid slabs, J. Opt. Soc. Am. 15 (4) (1998) 937–944.

[273] L. Wang, S. Jacques, Hybrid method of Monte Carlo simulation and diffusion theory for light reflectance by turbid media, J. Opt. Soc. Am. 10 (8) (1995) 1746–1752.

[274] L. Wang, S. Jacques, L. Zheng, MCML – Monte Carlo modelling of light transport in multi-layered tissues, Comput. Methods Programs Biomed. 47 (1995) 131–146.

[275] S. Williamson, H. Cummins, Light and Color in Nature and Art, John Wiley & Sons, New York, 1983.

[276] B. Wilson, G. Adam, A Monte Carlo model for the absorption and flux distributions of light in tissue, Med. Phys. 10 (1983) 824–830.

[277] A. Witt, Multiple scattering in reflection nebulae. I. a Monte Carlo approach, Astrophys. J. Suppl. Ser. 15 (1977) 1–6.

[278] R. Woodward, B. Cole, V.P. Wallace, R. Pye, D. Arnone, E. Linfield, M. Pepper, Terahertz pulse imaging in reflection geometry of human skin cancer and skin tissue, Phys. Med. Biol. 47 (2002) 3853–3863.

[279] G. Wyszecki, W. Stiles, Color Science: Concepts and Methods, Quantitative Data and Formulae, second ed., John Wiley & Sons, New York, 1982.

[280] M. Yamaguchi, M. Mitsui, Y. Murakami, H. Fukuda, N. Ohyama, Y. Kubota, Multispectral color imaging for dermatology: application in inflammatory and immunologic diseases, in: 13th Color Imaging Conference, Scottsdale, Arizona, 2005, pp. 52–58.

[281] A. Yaroslavsky, A. Priezzhev, J. Rodriguez, I. Yaroslavsky, H. Battarbee, Optics of blood, in: V. Tuchin (Ed.), Handbook of Optical Biomedical Diagnostics, SPIE Press, Bellingham, Washington, 2002, pp. 169–216.

[282] A. Yaroslavsky, S. Utz, S. Tatarintsev, V. Tuchin, Angular scattering properties of human epidermal layers, in: Human Vision and Electronic Imaging VI, SPIE, vol. 2100, Bellingham, Washington, 1994, pp. 38–41.

[283] E. Yeargers, L. Augenstein, UV spectral properties of phenylalanine powder, Biophys. J. 5 (1965) 687–696.

[284] G. Yoon, Absorption and scattering of laser light in biological media – mathematical modeling and methods for determining optical properties, Ph.D. thesis, University of Texas at Austin, Texas, 1988.

[285] G. Yoon, S. Prahl, A. Welch, Accuracies of the diffusion approximation and its similarity relations for laser irradiated biological media, Appl. Opt. 28 (12) (1989) 2250–2255.

[286] G. Yoon, A. Welch, M. Motamedi, M. van Gemert, Development and application of three-dimensional light distribution model for laser irradiated tissue, IEEE J. Quantum Electron. QE-23 (1987) 1721–1733.

[287] H. Zahouani, R. Vargiolu, Skin line morphology: tree and branches, in: P. Agache, P. Humbert (Eds.), Measuring the Skin, Springer-Verlag, Berlin, 2004, pp. 40–59.

[288] G. Zerlaut, T. Anderson, Multiple-integrating sphere spectrophotometer for measuring absolute spectral reflectance and transmittance, Appl. Opt. 20 (21) (1981) 3797–3804.

[289] G. Zonios, J. Bykowsky, N. Kollias, Skin melanin, hemoglobin, and light scattering properties can be quantitatively assessed *in vivo* using diffuse reflectance spectroscopy, J. Invest. Dermatol. 117 (6) (2001) 1452–1457.

Index

167